Battlefield Weapons Systems
& Technology Volume II

GUNS, MORTARS

& ROCKETS

Other Titles in the Battlefield Weapons Systems and Technology Series

General Editor: Colonel R G Lee OBE, Military Director of Studies at the Royal Military College of Science, Shrivenham, UK

This new series of course manuals is written by senior lecturing staff at RMCS, Shrivenham, one of the world's foremost institutions for military science and its application. It provides a clear and concise survey of the complex systems spectrum of modern ground warfare for officers-in-training and volunteer reserves throughout the English-speaking world.

For full details of these and future titles in the series, please contact your local Brassey's/Pergamon office

Related Titles of Interest

GUNS, MORTARS

& ROCKETS

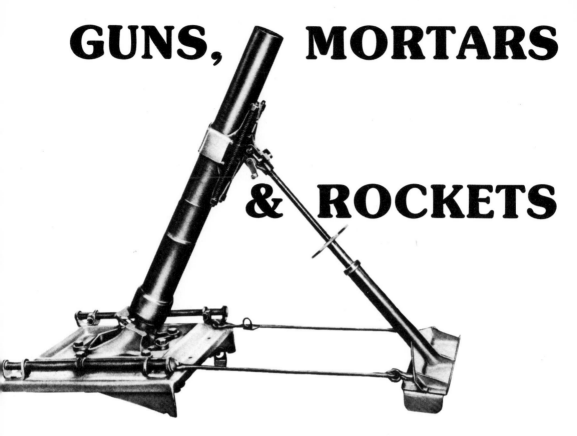

J W Ryan
Royal Military College of Science, Shrivenham, UK

BRASSEY'S PUBLISHERS LIMITED
a member of the Pergamon Group

OXFORD · NEW YORK · TORONTO · SYDNEY · PARIS · FRANKFURT

U.K.	BRASSEY'S PUBLISHERS LIMITED, a member of the Pergamon Group, Headington Hill Hall, Oxford OX3 0BW, England
U.S.A.	Pergamon Press Inc., Maxwell House, Fairview Park, Elmsford, New York 10523, U.S.A.
CANADA	Pergamon Press Canada Ltd., Suite 104, 150 Consumers Rd., Willowdale, Ontario M2J 1P9, Canada
AUSTRALIA	Pergamon Press (Aust.) Pty. Ltd., P.O. Box 544, Potts Point, N.S.W. 2011, Australia
FRANCE	Pergamon Press SARL, 24 rue des Ecoles, 75240 Paris, Cedex 05, France
FEDERAL REPUBLIC OF GERMANY	Pergamon Press GmbH, 6242 Kronberg-Taunus, Hammerweg 6, Federal Republic of Germany

First edition 1982

Library of Congress Cataloging in Publication Data

Ryan, J. W.
Guns, mortars & rockets.
(Battlefield weapons systems & technology ; v. 2)
Bibliography: p.
Includes index.
1. Ordnance. 2. Mortars (Ordnance) 3. Rocket
launchers (Ordnance) I. Title. II. Series.
UF520.R9   1982   623.4   82-3785
ISBN 0-08-028324-1 (Hardcover)   AACR2
ISBN 0-08-028325-X (Flexicover)

In order to make this volume available as economically and as rapidly as possible the author's typescript has been reproduced in its original form. This method unfortunately has its typographical limitations but it is hoped that they in no way distract the reader.

The views expressed in the book are those of the authors and not necessarily those of the Ministry of Defence of the United Kingdom.

Printed in Great Britain by A. Wheaton & Co. Ltd., Exeter

Preface

The Series

This series of books is written for those who wish to improve their knowledge of military weapons and equipment. It is equally relevant to professional soldiers, those involved in developing or producing military weapons or indeed anyone interested in the art of modern warfare.

All the texts are written in a way which assumes no mathematical knowledge and no more technical depth than would be gleaned from school days. It is intended that the books should be of particular interest to army officers who are studying for promotion examinations, furthering their knowledge at specialist arms schools or attending command and staff schools.

The authors of the books are all members of the staff of the Royal Military College of Science, Shrivenham, which is comprised of a unique blend of academic and military experts. They are not only leaders in the technology of their subjects, but are aware of what the military practitioner needs to know. It is difficult to imagine any group of persons more fitted to write about the application of technology to the battlefield.

Volume II

This volume is written for those who seek to improve their knowledge of guns, mortars and free-flight rockets. The scope of the book includes some historical background on the development of indirect fire weapons. This is followed by an examination of the basic requirements for artillery and a comparison of the contending systems. Guns, mortars and rockets are then covered separately, with emphasis on the design characteristics of each and comments on some of the available options. The book concludes with a look at the future and how the trends in technology will affect the employment and mix of indirect fire weapons fielded by armies of the future.

Shrivenham. December 1981. Geoffrey Lee

Acknowledgement

The author is grateful for the valuable advice and assistance given by the Department of Mechanical Engineering at the Royal Military College of Science, Shrivenham.

Shrivenham
September 1981   JWR

Contents

List of Illustrations

1.

Historical Background

INTRODUCTION

The history of artillery deserves far greater attention than is possible in a limited work such as this. Much of the history is largely irrelevant to the purpose of this volume, which is to describe technical aspects of modern artillery systems and the things that affect their design characteristics. Nevertheless it would be inappropriate to begin the consideration of present day techniques and equipment without first mentioning some of the major historical influences and events that have led to the equipment used by armies today.

THE ORIGINS

The origins of artillery are obscure. Ever since man decided that it was safer to inflict injury on an opponent from a distance there has been a gradual development of methods for hurling projectiles at the enemy. Some of the earliest devices included crude machines such as catapults, trebuchets and ballistas. The most significant difference between these weapons and modern artillery is that they were mechanical implements for the sudden release of energy to propel projectiles. By comparison, modern artillery weapons rely on the ability to harness the energy released by burning propellent to push the projectile out to the required range. Exactly when some of these weapons first appeared is difficult to pinpoint; however, large mechanical weapons for throwing projectiles were mentioned in the Old Testament.

Even by the time of Henry VIII the term 'artillery' was ascribed to a variety of weapons that could perhaps be regarded as more closely related to modern infantry weapons rather than artillery. For example, the charter bestowed by Henry VIII upon the Guylde of Saint George in 1537 charged them with becoming,

> "The overseers of the science of artillery .... to witt, long bowes, cross bowes and hand gonnes for the better encrease of the defence of our realme".

1

This charter is kept by the Honourable Artillery Company, an organisation that evolved from the Guylde. Clearly in those days artillery was the term given to almost any means of directing a projectile at an enemy. Nowadays the term is used to describe various forms of indirect fire systems providing fire support to armies in the field, especially guns, mortars and rockets. The term usually encompasses certain air defence systems, locating devices, as well as some long range anti-tank weapons; however, none of these will be covered in this book.

Rockets

The transition from one concept of artillery to the other took many centuries and was marked by a series of scientific discoveries and engineering developments. Rockets existed in China as early as the tenth century and crude rocket-propelled weapons used by the Arabs had been introduced to Europe by the thirteenth century. At this stage of the development of artillery, rockets began to give way to a more effective weapon, the cannon. The development of rockets remained dormant until more efficient applications of rockets were introduced in the early nineteenth century. These will be discussed later in this chapter.

Cannons

The earliest cannons probably had wrought iron barrels fashioned by lasking iron bars or staves around a white hot metal core and then arranging white hot, wrought iron hoops over the staves which compressed on cooling. The origin of the word barrel stems from this process. Some of the earliest examples were breech-loaded, for the manufacturing technique lent itself to this method of loading and obturation. The standard of obturation proved unsatisfactory and the breech loading approach was dropped. The purpose of the barrel in these early guns was simply to provide a vessel for the burning propellent, which at that time was black powder and to enable the projectile to be held and pointed in the right direction. In this regard they were truly the forerunners of modern guns. But it was the use of black powder as a propellent that was perhaps the most significant innovation, for without it the cannon would not have been introduced.

There are several conflicting claims as to who deserves the kudos for the invention of black powder or gunpowder. In most modern references it is accepted that the English friar Roger Bacon (C1214-1292) produced the earliest record of the ingredients that we can trace. Nonetheless most admit that it existed as a pyrotechnic for fireworks centuries before Bacon's time. The first cannons using gunpowder appeared in the fourteenth century and it was not until the nineteenth century that a better propellent charge, nitro-cellulose or gun cotton, was used. Throughout this period of approximately 500 years there were several other developments, the results of which are still clearly discernible in the design of modern guns and the techniques for their employment.

The Beginnings of Ballistics

During the fifteenth century cannons with trunnions began to appear. These were short stub axles fitted at the barrel's point of balance about which the elevation and depression of the piece was effected. Obviously gunners were becoming interested in being able to vary, quickly and easily, the range achieved by their weapons. It was about this time too that the Italian mathematician Niccol Tartaglia (1499-1577) advanced his theories in his work La Nova Scientia on the curved shape of trajectories or, as he described it, "the way of the pellet". Gallileo (1564-1642) later concluded that the flight path of a projectile was parabolic. Thus both men effectively debunked the belief held by many gunners that a projectile travelled in a straight line before dropping vertically to the earth. Much later the English mathematician and engineer Benjamin Robins (1707-1751) invented a device called the ballistic pendulum for use in determining muzzle velocity. Although this instrument has nowadays been replaced by the electric chronograph it still has a limited use for measuring recoil energy. Robins also suggested that the stability of a projectile in flight could be improved if it were rotated as it passed through the air. Unfortunately the manufacturing techniques available during his lifetime were unable to cope with the production of rifled cannons to prove his theories. Nevertheless, the application of ballistics had begun and well before this time new concepts for the tactical employment of artillery weapons were also being put into practice.

Changes in Tactics

Before the time of Gustavus Adolphus (1594-1632) artillery was comparatively immobile with its most common tasks being to demolish fortifications or to fire from prepared positions at an advancing enemy. He introduced a new dimension to artillery with the use of light, mobile 4 and 9 pounder demiculverins to support his troops. His artillery was organised to meet three distinct roles: siege, regimental and field artillery. Although his ideas proved to be successful he learnt the lesson in his experiments with tin-lined leather guns, that guns cannot be too light if they are to withstand firing stresses. Frederick II the Great (1711-1786) of Prussia followed his lead with his employment of artillery in the Seven Years War and extended the concept to include horse artillery. The eighteenth century saw the arrival of one of the most significant artillerists in history, Jean Baptiste de Gribeauval (1715-1789). His reforms as the French Inspector General of Artillery were the basis of the artillery triumphs of the Napoleonic era. He grouped artillery resources into coastal/garrison, siege and field. In his parlance the term "field" described 12 pounder weapons which were moved by horse drawn transport and supported by limber waggons hauling ammunition and spares. The United States Army adopted his system in 1809 and retained it until the 1840s. The improvement in techniques for the tactical employment of artillery were not accompanied by many significant developments in the design of guns.

Rifling

By the early nineteenth century artillery pieces still lacked many of the refinements apparent in modern weapons. A typical gun was muzzle loading, smooth bored and lacked any efficient mechanism to stop it careering backwards on firing. In the second half of the nineteenth century developments in gun design began to move swiftly. Before that time guns were able to fire from exposed positions on

advancing infantry with relative security down to ranges of about 400 metres. At
the Battle of Waterloo (1815) the opposing forces closed to about 200 or 300 metres
of each other before inflicting heavy casualties with small arms. Beyond those
ranges, however, guns could be used to some advantage. The invention of rifled
small arms made gunners vulnerable at much greater ranges and the logical extension of the innovation of rifling to artillery became apparant. The French and
the Prussians had rifles in quantity before 1850 but rifled field artillery was not
used until 1856 by the French in the Italian campaign.

Breech Loading Returns

As mentioned earlier, breech-loading artillery was used for the earliest cannons
but only because the manufacturing process demanded it and not because it was,
at that time, an effective means of providing obturation. Muzzle loading arrangements for cannons became the norm and held sway until manufacturing techniques
became good enough to overcome what had become age-old problems in obturation.
But like so many design innovations, the renewed demand for breech loading was
precipitated and hastened by another external influence: the arrival of the rifled
guns. Loading rifled guns from the muzzle presented practical problems, although many were produced and employed successfully. The breech loading
system had the advantage of being quicker in operation and safer on two counts.
Firstly, the gun detachments did not have work in exposed positions in front of
the guns during loading: and secondly, there was no danger of loading two projectiles and charges at once: a distinct possibility with muzzle loading weapons in the
heat of battle and one that remains with mortars today. By the late nineteenth
century breech-loading rifled guns were here to stay. These weapons were also
equipped with yet another long-awaited design improvement, an efficient method
of checking the recoil of the gun on firing. The amount of unrestrained recoil produced by old cannons must have been awesome and dangerous to say the least.
Before the introduction of recoil mechanisms, cannons on board ships were sometimes allowed to recoil across the middle of the deck and masts had to be protected with some form of cushioning. By 1807 on HMS Victory the advances made
were small with the cannons restrained by pulleys and weights. The detachments
were used to haul their weapons back into position after firing. As a criterion for
the number of men needed, the detachment size was determined by providing one
man for every 500 lbs of recoil. If we applied the same principle to modern
heavy guns the detachment size would be well in excess of 1000. Of course the
difficulties of coping with recoiling cannons on board ship were exacerbated by
problems of space. Nevertheless field gunners had good reason for wanting to
solve the same problem. Manhandling guns back into position after firing was
exhausting and degraded the rate of fire. Indeed, accounts of the Battle of
Waterloo relate that towards the end of the battle gunners were no longer able to
cope with the sheer physical effort involved. After many earlier attempts to produce an efficient recoil mechanism, the French produced their now famous M1897
75 mm field gun. The French 75 design included a hydro-pneumatic recoil
mechanism capable of performing two important functions. It absorbed the recoil forces acting on the gun during firing and, unlike other methods under examination at the time, it returned the barrel to its original position.

The design of the French 75 mm recoil mechanism incorporated a hydraulic brake which absorbed recoil by the action of forcing oil through a small orifice. Simultaneously the motion of recoil compressed air in a cylinder and once the recoil force had been spent the elasticity of the compressed air guided the barrel back into place. The recoil cycle, though controlled, was now fast enough to have the potential for appreciable increases in rates of fire; however, the ammunition was also a consideration in the question of rates of fire. For guns to be truly regarded as Quick-Firing by the gunners of that era, the ammunition needed to be loaded in one piece. The French 75 mm met this requirement with its fixed one-piece ammunition. Unlike modern guns using cartridge cases it had a screw breech not a sliding block. A detailed description of the different types of breech mechanism is given in Chapter 4.

The French 75 mm gun had a high muzzle velocity producing a flat trajectory which was considered desirable because it produced a wide spread of shrapnel at the target. Besides its innovative recoil system it also used a system of "abatage" or brake blocks positioned beneath its wheels. Although the French 75 mm was a pioneer of the modern field guns it was not without its faults compared with the guns that were to follow. It had a single trail and very limited traverse. Whenever the gun had to be moved through an arc of more than 3 degrees the trail had to be shifted and the brake blocks repositioned. The original French 75 mm guns did not have shields; however, these were added later.

Throughout history, advances in weapon design have rarely remained the province of any one nation for very long. The French 75 mm was no exception. Its design characteristics were quickly copied and refined by many nations to the extent that by World War I the French 75 mm could no longer be regarded any anything special.

Fig. 1.   French M1897 75 mm QF Gun

Quick Firing (QF)

The origins of the term QF are worthy of examination: it is a term still used today. It was derived from the extra speed in loading achieved by the use of propellent charges in brass cases instead of in cloth bags. The brass case enabled
the breech mechanism to be of much simpler design and easier to operate. On
firing, the brass case containing the charge provided the method of obturation.
Loading and firing was also facilitated by the fact that the brass case housed the
means of ignition for the charge. As mentioned earlier, for a fast rate of fire an
efficient recoil mechanism was the other essential component of the system.
When gunners of the late nineteenth century were progressing towards the development of a QF gun their efforts were directed at both these components of the
system. To them a gun without an efficient recoil mechanism could not be truly
regarded as QF even if it did use brass cartridge cases. The original intention
behind the use of the term has become somewhat distorted over the years. The
designation QF today refers to the method of obturation in which a metal cartridge
case (not necessarily brass) is employed to prevent the rearward escape of gases
developed by the burning propellent. The confusing thing for the uninitiated is
that a QF system is not necessarily quicker in terms of rate of fire than a breech
loading (BL) system: the term BL referring to another method of obturation in
which the gas seal is provided by the breech itself. Yet another source of mis-
understanding is that in both the QF and BL systems the gun is loaded through the
breech. Further explanation of methods of obturation, together with the breech
mechanisms that are employed with them, will be covered in Chapter 4.

The QF system was not universally regarded as a good thing, at least initially.
Resistance to change was strong as is often the case, especially during a period
of peace when any suggested radical departures from established practices are
more difficult to validate. One argument put forward was that metal cartridge
cases would represent a dead weight in the loads carried in ammunition waggons
and gun limbers. This criticism is still true today but was clearly more significant in the days of horse-drawn transport. Moreover the improved rates of fire
offered by QF systems was not regarded by some as essential because the existing rates seemed perfectly adequate except perhaps in circumstances where the
guns themselves were threatened by a fast moving force such as cavalry.

There was also much argument about whether the cartridge case of a QF system
should be fixed to the projectile or separated. Clearly fixed ammunition saved
time. The calibre of the weapon was, however, an important consideration and
it was generally agreed that there were handling limitations for fixed ammunition at calibres above about 105 mm because of the weight of the complete round.
A further consideration was the ability to vary the amount of propellent in a
separate cartridge case: a facility that is nowadays generally accepted as a
standard requirement for indirect fire gun systems.

The term QF should not be confused with "burst fire". Burst fire is a term
coined in recent years to describe the high rate of fire that can be achieved from
some modern artillery weapon systems. The definition of burst fire and the
rationale behind the requirement for it will be covered in later chapters.

Mortars

Mortars were evident amongst the first artillery weapons. The earliest examples
were clumsy, short-barrelled weapons firing metal balls. The term "mortar"
was probably derived from a chemist's mortar as the earlier versions resembled
this shape. Some were made of brass, some of latten and others were of bronze.
Usually they were cast and later versions were made of cast iron once the techniques for producing sufficiently high temperatures to melt and pour the ore were
developed. Even after the design and manufacturing techniques for guns began to
improve, mortars continued to be used. Their employment varied with calibre,
but essentially they were used to lob projectiles on top of the enemy or to breech
fortifications. The characteristic that distinguished mortars from guns was that
they were always fired at elevations of 800 mils or greater. It is this characteristic that still identifies a modern mortar from a gun or a howitzer, regardless of
whether the mortar is breech or muzzle loading, or whether the bore is smooth
or rifled.

Mortars were not restricted to use by armies as siege or defensive weapons.
They were also fielded in the coastal defence role and as naval weapons,
especially for shore bombardment. Until around the turn of this century, the development of mortars was comparatively slow. Often they were of larger calibre
than most orthodox mortars of today. Calibres of 13 inches and even greater
were commonplace, although the barrels were much shorter in terms of calibres
than modern mortars. Barrel lengths of less than three times the mortar's
calibre were the norm. The barrel was usually fixed to a wooden base for the
larger calibre versions. More often than not the elevation of these weapons was
fixed at about 800 mils and the range was adjusted by altering the amount of
charge. It was not until the middle of the nineteenth century that fixed charges
were used and the range varied by changing the elevation of the barrel. The concept of using mortars as light, mobile weapons was realised much earlier. Certainly the British used mortars as pack artillery in Nepal in the early nineteenth
century.

World War I saw the arrival of weapons that are perhaps more recognisable as
the antecedents of the mortars in use today. The siege mentality that developed
in the protracted trench warfare in Europe provided the impetus for the adaptation of an old siege weapon: the mortar. Unlike its predecessors the first examples of the new twentieth century generation of mortars were light, mobile and
very much infantry weapons used to supplement, or at times replace, the indirect
artillery support provided by guns. The Germans were the first to use these
small, short-ranged trench mortars or trench howitzers as they were sometimes
called and other armies were quick to follow suit.

A plethora of copies and improved versions soon appeared. Rifled and smooth
bore, breech loading and muzzle loading examples were produced together with
some rather bizarre weapons using centrifugal force and compressed air to
launch their bombs. Larger calibre mortars were produced and the British
Army had calibres up to 9.45 inch. Large calibre mortars were useful in the
static conditions that prevailed in trench warfare; however, their lack of mobility was the main reason why their use was discontinued after World War I.
Nevertheless the mortar had well and truly re-established itself as a standard

weapon in modern armies. A typical ground-mounted mortar today is very
similar to the original Stokes 3 inch mortar produced for the British Army in
1915. Their retention seems assured for the foreseeable future because of the
high rate of fire that can be produced at a reasonable range from a comparatively
light and inexpensive equipment.

Rockets

Rockets re-emerged as potent weapons of war during the nineteenth century.
British troops were subjected to effective rocket fire from Indian rocket systems
during the siege of Seringpatam in 1799 and this induced them to begin developing
systems of their own. William Congreve was given the task of investigating the
possible use of rockets by the Royal Artillery. Congreve was an officer in the
Hanoverian Army who had been attached to the Royal Laboratory at Woolwich in
England where his father was the Comptroller. Congreve later succeeded his
father in this appointment.

He developed rockets with ranges of several thousand metres. They were used
initially as ship-borne weapons during the attack on Boulogne harbour in 1806 and
similarly against Copenhagen in 1807. They were also used against the Americans
in the War of 1812 and were immortalised in the American National Anthem: "The
rockets' red glare" alluding to the British rocket bombardment of Fort McHenry.
Congreve published a book on rockets entitled "Details of the Rocket System" in
which he described the organisation, characteristics and tactics of his rocket
systems. Congreve's rockets came in different sizes from 6 to 42 pounders.
They were essentially a form of fixed ammunition fired without ordnance. The
rocket case was attached to a long stick to provide stability in flight. The rocket
payload varied. Some contained solid shot, some were incendiaries, while others
carried grenades or musket balls. Yet the most advantageous feature of
Congreve's rockets was, in his words, "The facility of firing a great number of
rounds in a short time, or even instantaneously, with small means". It is this
very feature that, amongst other things, enables present day rocket systems to
hold such an important place on the battlefield.

Rocket systems continued to be refined during the first half of the nineteenth
century although the improvements were primarily of an evolutionary nature.
Congreve rockets and Hale rockets (spin-stabilised) were both used during the
American Civil War, but their results were disappointing. Eventually the advantages offered by rifled barrels and recoil mechanisms firmly established guns as
the best available form of artillery, with the result that rocket systems dropped
out of favour. The development of artillery rocket systems languished for a time
until the period between World War I and World War II saw an upsurge in rocket
development. These developments, mainly in Germany, the United States and
the Soviet Union spawned the technology that was to lay the foundation, not just
for the free-flight artillery rocket systems of today, but also the many other
applications of rocketry that have transformed warfare.

SELF TEST QUESTIONS

QUESTION 1   In what way did Gustavus Adolphus contribute to the development
              of artillery?

              Answer   ....

                        ....

                        ....

QUESTION 2   Who invented gunpowder?

              Answer   ....

                        ....

QUESTION 3   Who was Jean Baptiste de Gribeauval and what did he contribute
              to the development of artillery?

              Answer   ....

                        ....

                        ....

QUESTION 4   What is the main characteristic that distinguishes a mortar
              from a gun?

              Answer   ....

QUESTION 5   What were the main reasons for the upsurge in the use of
              rockets in the nineteenth century?

              Answer   ....

                        ....

                        ....

                        ....

QUESTION 6   Were spin-stabilised rockets used in the nineteenth century or
              were they first introduced between World War I and II?

              Answer   ....

                        ....

QUESTION 7   Explain the term "Quick Firing" as it was originally used and
              the design innovations that led to its introduction.

Answer   . . . .

. . . .

. . . .

. . . .

. . . .

QUESTION 8   What were the main reasons for the decline in the use of rocket
systems in the second half of the nineteenth century?

Answer   . . . .

. . . .

QUESTION 9   Who was Tartaglia and what did he contribute to the development
of gunnery techniques?

Answer   . . . .

. . . .

. . . .

QUESTION 10   Why did breech loading guns disappear for several centuries
only to return to favour later as the normal type of gun?

Answer   . . . .

. . . .

. . . .

. . . .

QUESTION 11   What effect did the introduction of the rifled musket have on the
development of artillery?

Answer   . . . .

. . . .

. . . .

. . . .

QUESTION 12 Who was Benjamin Robins and what was his contribution
to the development of artillery?

Answer   . . . .

. . . .

. . . .

QUESTION 13 What is a "trench howitzer" and how did the term originate?

Answer   . . . .

. . . .

. . . .

ANSWERS ON PAGE 185

2.

Basic Requirements for Artillery Systems

INTRODUCTION

Modern artillery systems exist to provide 'indirect fire' onto targets. Whereas most artillery weapons can be used in the 'direct fire' role and have sighting systems to enable them to cope with such engagements, this role is primarily one reserved for infantry and armoured weapons. The requirements for indirect fire today demand a variety of effects at the target. The characteristics of the projectile used ultimately determines the effect achieved. Ammunition options will be discussed in outline in this Volume but for a detailed description of the different types of ammunition available see Volume III.

The design features of the delivery system for the projectile are influenced by the mass of the projectile as well as the range and accuracy to be achieved. Another important consideration is the nature of the operation for which the weapon is to be used. This, in turn, dictates the level of strategic, tactical and battlefield mobility; and the level of protection needed for the weapon and its detachment or crew.

In selecting the best artillery weapon for the task, be it a gun, mortar or rocket, the relative importance of weight of projectile, range and accuracy, mobility and protection, will often vary and even conflict. These conflicting requirements can never be fully satisfied in any one equipment. Consequently modern armies are equipped with a family of weapons to fulfil different primary indirect fire tasks. Equally the individual weapons themselves often mirror the need to reach a compromise in their design characteristics.

AMMUNITION

High explosive (HE), smoke and illuminating are still the most commonly used natures of ammunition. Whether this is likely to continue is debatable. New natures of ammunition such as cannon-launched guided projectiles and projectiles carrying scatterable sub-munitions could alter this balance in the future. The

likely impact of these improved munitions will be discussed in Chapter 8 and a
detailed description of their characteristics can also be found in Volume III.

High Explosive

The purpose of HE projectiles is to disperse high velocity fragments at the target.
The lethal effect of these fragments depends on their number, size and velocity.
This effect is accompanied by a destructive effect from the blast of the detonation
of the projectile. The latter effect, although most useful, can be regarded as
secondary. Indeed modern HE projectiles are designed for maximum lethality
against unprotected personnel and relatively soft-skinned vehicles and installa-
tions. The effects can be varied by the use of different methods of fuzing. These
methods can be grouped into three separate functions: airburst, groundburst, and
delay. HE projectiles fuzed for airburst produce a downward spread of fragments
enabling the neutralisation of troops behind vertical cover or in slit trenches.
They can also be employed effectively against troops in the open. Groundburst
fuzes function at the instant of impact producing an effective dispersion of fragments. The penetrative and cratering effects are small. Delay fuzes function
after impact, a delay of .05 of a second being typical. The result is a much
greater cratering effect which can be used to penetrate overhead protection and to
enhance the destructive blast effect on detonation. Clearly delay fuzes are more
effective against field defences and buildings and can be used to greater advantage
with heavier projectiles.

In deciding between the groundburst and airburst options for the attack of personnel the choice is not as simple. The optimum result will depend on the target
array, in particular whether the troops being engaged are erect or prone and indeed, if erect, whether they are likely to adopt the prone position as soon as the
engagement commences. The likelihood of immediate evasive action being taken
by troops under fire has been the driving force behind recent attempts to produce
guns with "burst" rates of fire. The term "burst fire" implies a capability to
fire a high number of rounds in the first 10-20 seconds of the engagement before
troops can lie down or take cover in trenches. In practice the first 4-5 seconds
are probably the most significant, hence the search for methods of improving
burst fire rates continues. The possibility of inflicting greater casualties with a
burst fire rate also means that a degree of lasting neutralisation may be achieved.
By comparison, slower rates of fire are more likely to have a much reduced
lasting neutralisation effect. Of course the need for burst fire has some attendant penalties. It usually requires some form of automation of the loading
sequence: certainly for calibres above 105 mm. In addition, it assumes that the
initial rounds of the engagement can be directed accurately onto the target without any adjustment or ranging.

Figures 2 and 3 show the relative effectiveness of height of burst for different
target arrays. The proximity of own troops to the fire must be considered
because the wider distribution of fragments when using airburst incurs the
penalty of an increase in the minimum distance to which own troops can approach
without taking unacceptable risks.

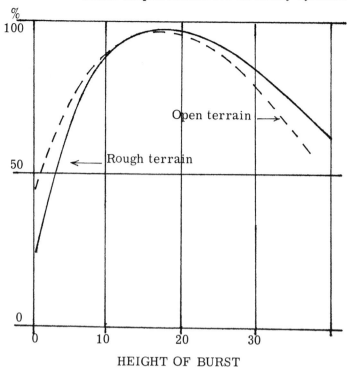

Fig. 2.   Effect of height of burst (troops prone)

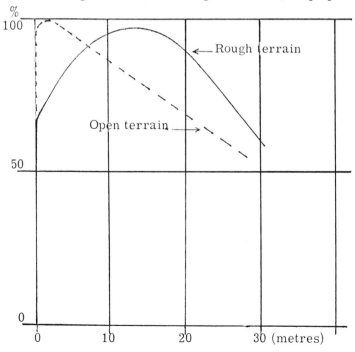

Fig. 3.   Effect of height of burst (troops erect)

Guns, Mortars and Rockets

Smoke

Smoke projectiles are used mainly to blind enemy observation and thus inhibit the
use of aimed direct fire and observed indirect fire. Its other application is to
screen from the opponent's view the movement of own troops. Further uses are
in marking targets for engagement by other weapon systems and for giving fire-
arranged signals. These applications are, however, very much secondary to
those of blinding and screening. The main disadvantage of smoke projectiles is
that their effects vary with meteorological conditions, especially wind. The
major advantage is that in suitable weather conditions they can blind or screen a
much larger area, for a longer period of time, than can be effectively neutralized
by the same number of HE projectiles. Figure 4 shows an example of dimensions
of the screen produced by a 105 mm base-ejection smoke projectile.

P   = Point of origin
H   = Point where height of cloud is effective
W   = Point where width of smoke cloud is effective
I   = Smoke ineffective
P-H = Distance to effective smoke height
P-W = Distance to effective smoke width
Note: Distances are for 105 mm BE Smoke and will vary with weather and
                                                                                ground.

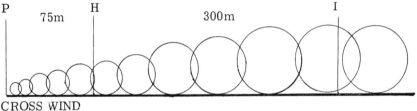

CROSS WIND

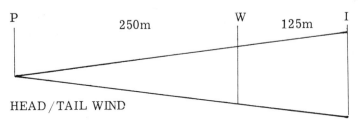

HEAD/TAIL WIND

Fig. 4.   Dimensions of 105 mm BE Smoke Screen

Other Natures

Other natures of ammunition commonly used include direct fire anti-tank projectiles, coloured flare projectiles and illuminating projectiles. These provide
artillery with the capability to perform additional, albeit very much secondary,
tasks to those performed by HE and smoke. In many conflicts since World War
II, fairly extensive use has been made of illuminating projectiles as an aid to
night observation. Recent improvements in other methods of providing observation at night such as image intensifiers, low light television and thermal imaging,
means that there is a trend towards carrying a much lower percentage of illuminating ammunition.

RANGE

The Need

The quest for greater range has continued ever since artillery first came into being. Some of the reasons for this are obvious, others are less apparent. The obvious reasons are that greater range permits the engagement of targets over a much larger area and also increases the possibility of concentrating the fire of more guns on a given target. The other reasons are that the greater the range capability of his weapons the easier it is for a commander to position his artillery beyond the range of enemy indirect fire weapons and the less may be his need for mobility. A long range becomes particularly important when an army is advancing or withdrawing. In these phases of war the extended distances over which artillery support may be needed often necessitates rapid re-deployment of artillery resources. Because artillery on the move is useless it follows that a good way of decreasing the time out of action is to improve the range of artillery. As a yardstick it can be said that for a given artillery role, such as the provisions of close support to own troops, nations will continue to endeavour to out-range the opposition. Yet there are penalties to be paid for increased range.

Increasing Range

There are some methods available to improve maximum range without altering the equipment used to deliver the projectile. There are two approaches to this: the first is to improve the 'ballistic coefficient' of the projectile and thus improve its ability to pass through the air more easily. The second is to give the projectile some form of post-firing boost during flight. These methods will not be considered further although some important implications of their introductions will be covered in Chapter 8. For a given calibre and mass of projectile an increase in range usually results in a heavier gun because of the need to withstand the added firing stresses from the larger propellent charge needed to produce a high enough muzzle velocity to achieve the range. Even if the additional weight is acceptable there are further problems. The major ones are range coverage, 'accuracy' and 'consistency'.

Range Coverage

A gun with a fixed propellent charge producing a high muzzle velocity may well be able to achieve a given range; but for practical purposes the shape of the trajectory may be too flat particularly at the shorter ranges. One disadvantage of this characteristic is that in mountainous terrain the trajectory may be too flat to reach down behind hills or into valleys. Gunners have been faced with this problem for some time. At the battle of Waterloo (1815) Wellington's troops were positioned along the line of heights covering the road to Brussels and sheltered from French artillery by being on the reverse slope. The method used to overcome this problem is to employ a variable propellent charge system producing different muzzle velocities for different range brackets. Hence for a given range, except for maximum or near maximum range, different muzzle velocities producing different trajectories can be used.   The number of charges needed to give a good selection of trajectories generally depends on the maximum range. The U.S. 105 mm M2A2 Howitzer uses 7 charges to produce the full range cover- age out to 11,000 metres and the British FH70 uses 9 charges for range coverage out to 24,000 metres.   Figure 5 shows an example of range coverage.

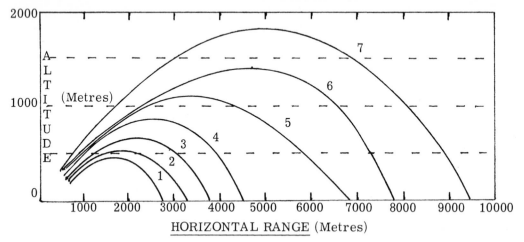

Note: Diagram shows ranges achieved
by 105 mm L5 Pack Howitzer
firing charges 1 to 7 at an
elevation of 600

Fig. 5.   Range coverage

Variable charge systems are an added complication in gun drills and if an auto- matic loading system is used the greater the number of charges the greater the mechanical complexity of the system.   This in turn could present problems in reliability and cost.

ACCURACY AND CONSISTENCY

Definitions

For the purposes of gunnery the terms accuracy and consistency are defined as follows. 'Accuracy' is a measure of the precision with which the 'mean point of impact' (MPI) of a group of rounds can be placed on a target. 'Consistency' is a measure of spread of rounds around the MPI when the rounds are fired from a gun at the same elevation. When the dispersion of rounds about the MPI is large this consistency is said to be poor. Accuracy is a function of the overall system and is affected by many sources of errors and mistakes.   They include the survey of guns, the accuracy with which the target location has been determined; and other sources of inaccuracy inherent in the ammunition, instruments and correc- tions applied for external conditions. Consistency is affected by round to round variations in a number of things including muzzle velocity, ballistic coefficient,

meteorological conditions, 'laying', 'ramming', and 'wear' of the gun. It is possible for artillery fire to be consistent but inaccurate (and vice versa). On the
other hand, if the projectiles from all guns landed at the centre of the target on
every occasion then both accuracy and consistency could be regarded as perfect.
But even this result would be less than ideal. Certainly some degree of dispersion around the MPI of projectiles delivered from a gun produces an effective
and desirable area coverage, assuming that a sufficient number of rounds is
fired. For practical purposes artillery forward observers apply pre-determined
data from 'firing tables' or 'range tables' that predict the likely length and breadth
of the dispersion about the MPI. The dispersion plus and minus of the MPI along
the line gun - target is the most critical. An example of a typical spread is shown
in Figure 6.

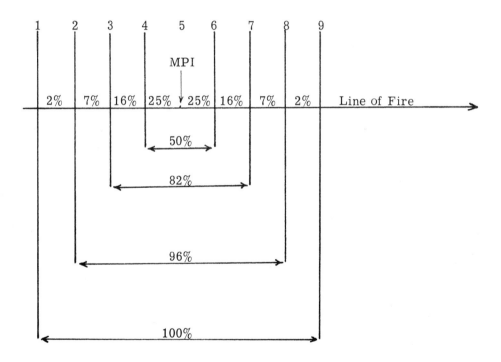

Note: 1.   50% of rounds fired will probably fall between lines 4 and 6,
             82% between lines 3 and 7 and so on.

        2.   An alternative explanation of the diagram is that if a single
             round is fired it has a 50% chance of falling between lines 4
             and 6.

        3.   The distance between adjacent vertical lines is equal and
             for a given charge varies with range.

Fig. 6.   Dispersion

As previously mentioned, some degree of dispersion is desirable as long as it is predictable, which it is. Guns are rarely employed singly in the indirect fire role when firing conventional HE projectiles. It is common practice to use a minimum of six or eight guns when engaging a target. Obviously the greater the number of guns that can concentrate on a target simultaneously the more effective the fire. Whatever the number of guns used to engage a target the dispersion figures for line and range are the basis for calculating the area that will be covered by the effects of the projectiles arriving at the target.

Another factor affecting the distribution of rounds at the target is the spread of the guns themselves at the gun position. Since it is normal practice to fire guns from the same location with their lines of fire parallel, the distance between individual guns will be reflected in the displacement of their MPIs at the target. If necessary, allowances can be made for this when computing gun data so that the lines of fire converge and the MPIs of all guns are directed at the centre of the target. For a given range the dispersion along the line of fire is greater than the dispersion across the line of fire. Furthermore, the degree of dispersion will generally increase with range for a given charge. For example, the 105 mm L5 Pack Howitzer firing at top charge will spread its rounds over an area of 208 metres along the line of fire and 24 metres across the line of fire when firing at a range of 5000 metres. At a range of 10000 metres the area will be 272 x 48 metres. The added dispersion may sometimes be desirable; however, on occasions it can be inconvenient because own troops will not be able to move in as close to the MPI.

The accuracy of the overall system is a different matter. Ideally it should be perfect so that the MPI of a group of rounds can be placed as close to the centre of the target as possible without having to adjust the MPI. The nature of many of the components of the system affecting accuracy is such that at greater ranges a high degree of accuracy may be more difficult to achieve. For example, the longer the range of the equipment the greater the area for which accurate corrections for weather must be deduced. Any angular difference between the correct line of fire between gun and target and the actual line along which the projectile is delivered is manifested as a displacement between the centre of the target and the MPI. Obviously this displacement will increase with range.

MOBILITY

General Considerations

Ever since the time of Gustavus Adolphus (1594-1632) the importance of mobility has been realised. Mobility remains one of the prime considerations in the overall design balance of artillery because, as we shall see, a given level of mobility often demands a compromise on other basic weapon characteristics.

The nature of the operations in which the weapons will be employed has an important bearing on the level of mobility, as does the level of mobile of the other components of the force. One often hears the claim that modern artillery must have
mobility on the battlefield comparable to that of the supported arms-infantry and
armour. This can be somewhat misleading and depends on exactly what is meant
by comparable. Furthermore, in certain scenarios one can advance a strong
argument for the mobility of artillery to be better than that of the supported arm.

For any future operations in North West Europe involving NATO and WARSAW
PACT forces the opposing sides will place much greater reliance on motorised
transport and hence mobility than before. The trend is towards greater use of
'self-propelled' (SP) artillery and this makes good sense. Nevertheless, there
are certain aspects of mobility required of tanks and even mechanised infantry
combat vehicles that simply are not required in an SP. For example, the agility
demanded from a tank so that it can accelerate quickly from fire position to fire
position and make sudden sharp changes in direction requires a far more powerful engine and more efficient suspension and transmission systems than is required for an SP.

At the other end of the spectrum, future operations in difficult terrain may demand a much greater emphasis on dismounted infantry. In these circumstances
it may well be that the bulk of the transport resources will be reserved for
logistic support and the rapid redeployment of artillery resources. In other
words the artillery may be inherently more mobile than the supported arm in
terms of its ability to redeploy quickly over a long distance.

Design Factors Affecting Mobility

In addition to the nature of the task some of the more specific factors that will
influence the inherent mobility of artillery are the mass of the projectile, the
range at which it is to be delivered and the degree of protection afforded to the
detachment. The more powerful propellent charge needed to produce increased
range requires that the weapon be heavier so that it can withstand firing stresses
and still remain stable. Rockets provide an exception to the rule and this topic
will be covered in Chapters 3 and 7. Similarly, for a given range any requirement to deliver heavier projectiles will also lead to an increase in the weight of
the equipment to cope with the increased muzzle energy needed. Although mobility can be enhanced by certain design innovations such as the employment of
muzzle brakes to reduce recoil energy and the use of light alloys in carriage construction, the weight penalty for a given range and mass of projectile is difficult
to escape. Muzzle brakes produce blast overpressures that can be injurious to
the detachment and any reduction in weight by the use of light alloys cannot be
taken to the extent where the equipment becomes so light that it is unstable when
fired.

PROTECTION

The main threats to modern artillery weapons come from enemy indirect fire weapons and ground-attack aircraft. There are four main techniques used in the employment of artillery in the field to counteract these threats: dispersion; camouflage and concealment; digging; and mobility. The last of these has already been discussed. The others are largely a function of tactical employment and deployment procedures adopted and will not be discussed further. Another method is to incorporate some form of protection in the design of the equipment and this approach will now be discussed.

Shields

The earliest attempts to provide protection involved the use of shields for towed guns to give some form of ballistic protection, at least in one direction. The appearance of shields on guns followed closely behind the arrival of improved small arms on the battlefield. They were certainly a useful addition to guns in the days when artillery was deployed in more exposed positions in the battlefield. Gun shields have been retained even after gunnery techniques progressed to the stage where artillery was deployed to provide indirect fire from hidden, dispersed locations. The U.S. 105 mm M2A2 Howitzer, which is still in service in many armies, has shields, see Fig. 7. No doubt the rationale for their retention was based on three reasons. Firstly, they afforded a limited measure of protection against the effects of counter battery fire. Secondly, field guns on occasions have been pressed into service as direct fire anti-tank weapons, particularly during World War II and in such situations shields are highly desirable. Thirdly, gun shields can provide a limited degree of protection for some members of the detachment against damage to hearing as a result of blast overpressures produced on firing. Nevertheless, the advent of the guided missile in its various forms as a comparatively light, mobile, anti-tank weapon makes the possibility of guns being used in this role extremely remote. Additionally, air portability requirements for modern towed guns demand that weight is kept to an absolute minimum. It is not surprising, therefore, that the trend is to exclude gun shields in modern towed equipments.

Protection for Towed Systems

A towed gun is comparatively robust, with the most vulnerable parts of the equipment being the recoil mechanism and the sights. Of these the recoil mechanism presents the biggest problem because the replacement of a recoil mechanism is a major task. Although the design of many towed guns affords some protection for the recoil mechanism by virtue of its positioning beneath the barrel, in other design configurations it is mounted in an exposed position above the barrel. Figure 8 shows an American 105 mm M2A2 Howitzer with the upper recoil mechanism on top of the barrel. French gunners at the Battle of Dien Bien Phu (1953) were subjected to intense counter battery fire from Viet Minh artillery and mortars. The extent of the counter battery fire was such that the French-manned M2A2 105 mm Howitzer sometimes had to be dug out from beneath the earth and debris thrown up by exploding projectiles. The Howitzer generally proved to be invulnerable except to direct hits or very near misses; however the exception to this was the upper recoil mechanisms which seemed to be relatively easily hit and pierced by fragments.

LEFT AUXILIARY SHIELD

TOP LEFT FLAP

TOP RIGHT FLAP

UPPER LEFT SHIELD

UPPER RIGHT SHIELD

RIGHT AUXILIARY SHIELD

AUXILIARY SHIELD TIE PLATE

LOWER LEFT SHIELD

LOWER RIGHT SHIELD   BOTTOM FLAP

Fig. 7.   M2A2 Gun Shields

Fig. 8.   M2A2 Upper recoil mechanism

Interestingly, the towed gun most recently produced by the United States, the
155 mm M198, has perpetuated what is potentially a design weakness by having its
recoil mechanism positioned in a similar fashion. The potential danger, however,
has been recognised and the recoil arrangements have been protected with a ballistic shield.

A more ambitious attempt at protection for towed guns was the British Garrington
Gun shown in Fig. 9. This equipment was the result of a study done by Garrington
UK in the 1950s. The aim of the study was to produce a carriage on which to
mount an American 112 mm ordnance. The 112 mm ordnance was discontinued
and an 88 mm calibre ordnance was developed with a view to replacing the 25
pounder with a new equipment firing a redesigned projectile.

The Garrington Gun's most unusual feature was its overhead box trail which supported a thermal shield. Although the shield was designed primarily against the
thermal effects of a nuclear explosion, it also provided a degree of ballistic protection. The concept was short-lived, however, and the Garrington Gun was never
introduced into service, if only because a standardisation agreement was reached
on 105 mm calibre. Despite the fact that several towed guns have been designed
and introduced into service since the time of the Garrington Gun, there have been
no further attempts to produce a level of detachment protection similar to that of
the Garrington Gun. The Garrington Gun had a range of 15,500 metres with a
21 pound projectile. The equipment weight was 2270 kilograms and it had 6400
mils traverse and a high angle capability.

Fig. 9.   Garrington Gun (88 mm)

Self-Propelled Guns

SP guns offer the best scope for providing protection as part of the overall design of the equipment.   Most of the early SP guns were originally the result of attempts to provide increased mobility.   They came into their own in World War II, although some 155 mm SP guns were fielded by the French in World War I.   It was the rapid development of mobile armoured warfare in World War II that provided the impetus for the proliferation of SPs on the battlefield.   In many cases designers of early examples of SPs opted for the easy way out by simply removing the wheels and trails of towed guns and mounting them on existing tank, personnel-carrier or even truck chassis.   Sometimes it worked, as with the successful combination of the Canadian Ram Tank and the British 25-pounder gun to form the Sexton SP.   Sometimes it did not, as with the Lloyd 25-pounder: an unhappy combination of the same gun and a personnel carrier.   In many cases the protection varied from rather rudimentary ballistic protection to almost none at all. Even today some self-propelled guns in service are without all round protection for the detachment.

Fig. 10.   175 mm SP Gun

The latest SP guns are indicative of the increased importance placed on protection. Typically, they provide protection from a near burst of a projectile of
around 155 mm calibre and small arms fired at point blank range. The British/
German/Italian 155 mm SP70 will provide this sort of protection. The weight
penalty incurred for this level of protection is significant. By way of comparison
SP70 will weigh more than 40 tonnes and the American M109 155 mm SP, still in
service but introduced in the 1960s, weighs 24 tonnes. The search for better protection naturally has attendant penalties in terms of mobility too, especially strategic mobility.

Fig. 11.   155 mm M109 SP Guns

SELF TEST QUESTIONS

QUESTION 1 Why are armies equipped with a family of indirect fire weapons?

Answer ....

....

....

QUESTION 2 Why is there a trend towards carrying a much lower percentage of illuminating ammunition?

Answer ....

....

QUESTION 3 What are the two approaches to increasing range of guns by improving the ammunition?

Answer ....

....

QUESTION 4 Explain the term "range coverage".

Answer ....

....

....

....

QUESTION 5 Define the terms "accuracy" and "consistency".

Answer ....

....

....

....

....

....

QUESTION 6 List the advantages of having shields on towed guns and explain why the trend is towards not fitting shields.

Answer   . . . .

. . . .

. . . .

. . . .

QUESTION 7   Explain the term "burst fire" and reasons for the importance
placed on this capability in the design of artillery weapon systems.

Answer   . . . .

. . . .

. . . .

. . . .

QUESTION 8   List two of the characteristics of guns that may degrade its
mobility.

Answer   . . . .

. . . .

QUESTION 9   What methods can be used to provide protection for guns from
counter battery fire and what penalties does their use impose?

Answer   . . . .

. . . .

. . . .

. . . .

. . . .

ANSWERS ON PAGE 186

3.

Delivery Systems

INTRODUCTION

As mentioned in the previous Chapter, the choice of a weapon system to provide indirect fire support will be influenced by the nature of the task, the terrain, the effect required at the target and the relative importance given to range, mobility and protection. In modern armies the general consensus seems to be that no single form of indirect fire system is adequate for all contingencies; thus a combination of guns, mortars and rockets is used. The main characteristics of these systems will now be discussed and a comparison made of the suitability of each to fulfil certain tasks.

THE TASKS

For simplicity the tasks will be restricted to two. The definitions used here for these tasks may not necessarily match those adopted in all armies or indeed any for that matter. Nonetheless, they are sufficiently descriptive for the purposes of this volume; in particular as a basis for comparing guns, mortars and rockets. The first task is Close Support which is the artillery support given to infantry and armour to help them defend and seize ground, manoeuvre with impunity, but at the same time prevent the enemy from doing likewise. It remains an important task because the direct fire weapons of infantry and armour can be easily neutralised by indirect fire weapons. The second is Depth Support which is artillery fire applied in depth against enemy artillery, as well as infantry and armour beyond the immediate battle area.

GUNS

General Characteristics

Modern guns are well suited to Close Support tasks in terms of range and their multiple charge systems produce good range coverage and hence a wide selection

of trajectories. Most Western armies have a mixture of calibres, predominantly
105 mm and 155 mm with a marked tendency towards the latter in the last decade.
The lethality of these weapons is well-proven against troops in the open but
against armoured personnel carriers and, in particular, against tanks their effect
is much more limited. Their accuracy and consistency also enables them to compare favourably with other contending systems for Close Support. The need for
quick response to calls for fire and a high rate of fire can also be satisfied; however, high rates of fire are difficult to maintain over long periods with larger
calibre weapons without mechanical handling and loading arrangements at the
weapon. For Depth Support tasks the general characteristics already mentioned
still apply, except that the need for greater range and heavier projectiles becomes
more critical.

It is difficult to find a gun that is the perfect answer to both the Close and Depth
Support tasks. Although some recent examples of 155 mm weapons with ranges
out to around 30 kilometres are a good compromise, both the range and projectile weight are still not good enough. By comparison the existing large calibre
equipments such as the American 8 inch and 175 mm equipments, despite their
heavier projectiles, suffer from poor rates of fire and inadequate range. The
175 mm has a range of 32.8 kilometres and the 8 inch (unmodified) a range of
16.7 kilometres. This is not to say that there is no scope for the improvement of
heavier calibre guns to alleviate these problems. After all, 155 mm systems
have recently been improved to produce dramatic increases in range. For
example 155 mm guns of World War II vintage had ranges of around 15 kilometres,
whereas the latest examples of the same calibre have ranges of 24 kilometres with
conventional projectiles. Notwithstanding the fact that there may be scope for the
improvement of large calibre guns, the general trend seems to be towards other
solutions, in particular rockets. The final choice of a gun for a given task also
necessitates a comparative evaluation of SP and towed concepts.

SP Versus Towed

Perhaps the best way to compare SP and towed guns is to examine two contenders
of the same calibre firing the same ammunition to the same maximum range. In
this way the relative merits of the two design concepts can be assessed more
accurately. The British/German/Italian FH70 155 mm towed Howitzer and
155 mm SP70 provide a good basis for comparison even though SP70, at the time
of writing, is still at the prototype stage of its development. Both of these equipments are shown on the next page. The fact that they represent the combined
efforts of three nations also provides a clear perspective of the general trends in
gun design: trends that, incidentally, have since been followed by other Western
nations. The first impression may be that if the consortium that designed these
equipments identified a need for towed and SP systems of the same calibre, then
the conclusion is that both types are needed. One should keep in mind, however,
that FH70 was the result of much earlier assessments of the requirement and
that perhaps the same conditions may not apply today or in the future. A further
consideration is that both weapons have been produced primarily for use in
Europe. Therefore their characteristics and indeed their relative suitability
could have been different if they had been designed mainly for use in other parts
of the world.

Fig. 12.   FH70 155 mm

Fig. 13.   SP70 Phase B 155 mm (artist's impression)

The table below lists the important characteristics of these weapons.

Characteristics	FH70	SP70
Range	24 kilometres 30 kilometres (with rocket assistance)	24 kilometres 30 kilometres (with rocket assistance)
Weight of Projectile	43.5 kilograms	43.5 kilograms
Weight of Equipment	9 tonnes (approximately)	45 tonnes (approximately)
Ballistic Protection	No protection for detachment.	Can withstand near burst of 152 mm calibre.
Strategic Mobility	Air transportable.	Not air transportable.
Tactical and Battle-field Mobility	Transportable by medium lift helicopter equipped with an auxiliary power unit for short moves.	Self propelled.
Rate of Fire	Burst fire rate 3 rounds in 10 seconds using flick rammer. Maximum rate 6 per minute. Sustained rate 2 per minute.	Burst fire rate 3 rounds in 10 seconds using autoloader. Maximum rate 6 per minute. Sustained rate 2 per minute.
NBC Protection	Nil	Collective protection for crew.
Ammunition Re-supply	Both systems rely on the same logistic resupply system although SP70 can carry a limited number of rounds inside the vehicle.	

Examination of these characteristics indicates that the significant differences between the towed and SP solutions are in mobility (especially strategic mobility), protection and cost. In general terms this is valid for all towed and SP guns of the same calibre, even though the magnitude of the differences will vary if other specific examples are used as a basis for comparison. For example, if the American M198 155 mm Howitzer were compared with SP70 the variation in mobility and rate of fire would be even more marked in favour of SP70. The choice between SP and towed is by no means straightforward. Towed guns still have a place on the battlefield, especially where deployment by air and cost remain significant.

MORTARS

General Features

Unlike guns, modern mortars still bear a close resemblance to some of the
earliest forms of artillery weapons. Conventional mortars such as the one in
Fig. 14 retain three important design features that have since been rejected in the
design of guns. The first is that they do not have recoil mechanisms, the main
recoil force being transmitted to the ground through the base plate; the second
is that they are the only surviving muzzle-loading indirect fire weapons; and the
third is that they have smooth bores. The other important characteristic of conventional mortars is that the elevations at which they can fire are restricted to
angles above 45°. This means that mortars cannot be used in the direct fire
role.

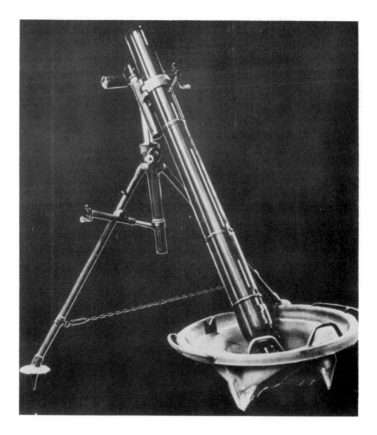

Fig. 14.   Conventional ground-mounted mortar

Some examples of mortars with rifled bores and breech loading mechanisms have been produced; however, they can be regarded as unorthodox versions of this type of weapon. The main disadvantages of these unorthodox design innovations for mortars are that they tend to degrade their rate of fire and their inherent simplicity in design. As we shall see these are two of the important characteristics of mortars and ones that should not be discarded lightly.

Fig. 15.   Breech loading mortar

Mobility of Mortars

The mobility of mortars compares more than favourably with other artillery weapon systems. The main components of a mortar are a base plate or some other rigid mounting to absorb firing stresses, a barrel and some form of support for the barrel. They are easy to assemble, fire, disassemble and maintain. Additionally, most mortars can be manhandled, even over long distances if required. This means that in very difficult terrain, mortars may be the only indirect fire weapons that can be deployed. Because of these features they are admirably suited to the Close Support of infantry. Indeed in many armies they are manned entirely by infantrymen; although in the past, but to a lesser extent today, they have sometimes been manned by gunners. It is important, however, to remember that for a given target effect the strain placed on the ammunition resupply system is comparable to that incurred by the use of guns. This is a point that is often overlooked by those who extol the mobility virtues of mortars. Because the successful re-supply of ammunition is often dictated by the mobility of the transport used for the task, this dimension of the overall system's mobility should be remembered when comparing it to other systems.

Mortar Ammunition

The characteristics of mortar ammunition have some significant differences to those of gun ammunition. Because mortar bombs are fired from smooth bores they cannot easily be spin stabilised; therefore, stability of mortar bombs in flight is achieved by the use of fins fitted towards the tail of the bomb. The effect is to increase the surface area behind the centre of mass so that when the axis of the bomb deviates from the trajectory the centre of air pressure acting on the bomb will move to the large area presented by the fins and tends to push the axis of the bomb back towards the trajectory.

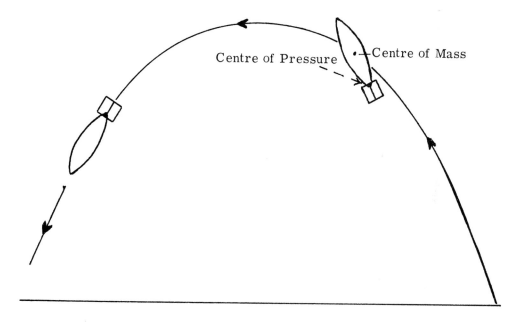

Centre of Pressure   Centre of Mass

Fig. 16.   Stability of mortar bombs

Although spin stabilisation is possible for mortars, muzzle loading then becomes complicated. The problem is to ensure that, when loaded, the bomb has enough downward velocity for the cartridge system at the tail of the bomb to function on the fixed firing pin at the base of the barrel. A solution is to provide a separate firing mechanism to be operated after the loader has moved clear of the muzzle. The disadvantages include an increase in mechanical complexity of the weapon and a lower rate of fire.

Lethality

Conventional mortars have sub-sonic muzzle velocities; therefore, mortar bombs are subjected to much lower pressures in the barrel than projectiles in a gun. These lower pressures permit the outer casing of mortar bombs to be thinner than in gun projectiles. This means that a high percentage of explosive filling can be included in the bomb. Furthermore, because of the lower pressures, the

designers of mortar bombs are not as constrained by considerations of strength
as is the case with designers of gun projectiles. Consequently, there is the
choice of cheaper bomb casing materials and/or materials with better fragmenting qualities than those used in projectiles for guns. The comparative lethality
of mortar fire is also enhanced by the steep angle of descent at the target and the
low remaining velocity of the bomb. These characteristics ensure a good spread
of fragments and also make mortars ideal for delivering chemical payloads.

The main disadvantages of mortar ammunition is its poor effect against hard targets. HEAT projectiles have been made; however the stand-off distances required for a hollow charge tend to result in the bomb's centre of gravity being
further back than is ideal for stability. A further disadvantage is that the tapered
shape of a mortar bomb does not make it an ideal carrier for loads such as flares
and smoke cannisters. Some of the other disadvantages traditionally ascribed to
mortar ammunition, such as difficulty in providing obturation and large tolerances
in the manufacture of bombs have been largely overcome in modern ammunition
production techniques, albeit at a cost.

Range

Mortars have comparatively poor maximum range. Furthermore, the scope for
improving the range of mortars is extremely limited. Rocket assistance can be
given as, for example, has already been done with the bomb for the 120 mm
Hotchkiss Brandt rifled mortar. But this expedient only results in a maximum
range of 13 kilometres. Furthermore, there is a loss of payload to accommodate
the rocket assistance. As a yardstick, rocket assistance is probably uneconomical below about 105 mm calibre because of the payload penalty. Even in larger
calibres this loss could be around 50%. Further, as with rocket-assisted gun
ammunition there are added difficulties in achieving an acceptable level of consistency in dispersion at the target.

Two of the characteristics of conventional mortars that are the basis of many of
their design advantages, smooth bores and low chamber pressures, are the
very factors that inhibit the likelihood of any significant increase in range. Theoretically a high chamber pressure generating supersonic velocities could produce a significant extension of range; however, the conventional fin-stabilised
bomb would be unable to cope with such velocities. An improved bomb capable
of use at these velocities may well be possible but the basic contours of the conventional bomb would have to be drastically altered. It would have to be more
streamlined and the size of the fins would need to be much larger to perform
their function efficiently. Additionally, the basic components of the mortar would
necessarily have to be made much stronger with a resultant weight penalty. The
outer casing of the bomb would have to be thicker to withstand the larger barrel
pressures with an attendant loss in payload. All in all the result would be a
design compromise; a compromise that would render the mortar system more
akin to a light gun in terms of cost and complexity. Hence mortars remain comparatively short ranged weapons: a characteristic that makes them unsuitable
for Depth Support tasks. If one assumes that a mortar firing supersonic bombs
is not a practical proposition, then the question arises as to their range limits
at subsonic velocities.

Transonic velocities (around 340 metres per second) pose difficulties if an acceptable level of consistency is to be retained. It is normal practice for designers to avoid this velocity bracket because of the uncertain ballistic performance produced. Consequently, muzzle velocities in excess of about 310 metres per second are unusual. The maximum range achievable in a vacuum at the optimum elevation of 45^0 is less than 12 kilometres. In reality the range would be much less because of losses caused by drag. We are therefore left with a maximum range of about 9 kilometres for a heavy mortar and considerably less for smaller calibres. Nevertheless we should not forget that despite the limited range the rates of fire achievable are excellent.

Rate of Fire

The limit to the rate of fire of a mortar is dictated, in theory at least, by the time between the action of dropping the bomb down the muzzle and the departure of the bomb from the mortar. Some claims put this rate as high as more than 30 bombs per minute but obviously crew fatigue would eventually degrade very high rates. Clearly several things such as: the calibre of the mortar, the size of the crew, and the length of the bomb and the barrel, will affect both the maximum and sustained rates. Suffice to say that in comparison guns can only begin to match mortars in rates of fire by the use of complex and expensive automatic loading systems.

Accuracy and Consistency

Although modern mortar systems have improved in terms of accuracy and consistency compared to their predecessors, gun systems still have the edge in this regard. Any difference in consistency is really not of great significance; however, the accuracy of the systems is another matter. Although it is technically possible to make the necessary corrections for conditions affecting accuracy, in most cases this is not done to the same extent as for guns, if only to preserve one of the prime advantages of the system - simplicity.

Protection of Mortars

A level of protection for mortars from counter battery fire can be achieved by: frequent movement; digging and concealment; and to a lesser extent dispersion. The high trajectory of a mortar bomb and its finned bomb make it most vulnerable to detection by mortar locating radars. A further disadvantage is that mortars normally have to fire at least two rounds at a high elevation and charge to ensure that the baseplate has been pushed down firmly enough into the earth to form a stable platform for subsequent firing. This process is called "bedding-in" and besides having disadvantage of being wasteful of ammunition, it could compromise the location of the mortars. Frequent redeployment is the best means available to avoid counter fire; but like any other indirect fire system mortars are useless when not deployed and ready to fire. Any design innovations, therefore, that degrade their mobility or their speed into and out of action perforce degrade their best means of protection. Yet another argument to keep mortars light and simple.

Because they are relatively small, dismounted mortars are easy to dig-in and conceal compared with other systems. In addition, because they always fire in high angle some measure of protection can be achieved by deployment behind steep cover. Complete overhead protection for the crew is difficult to provide because the loader needs ready access to the muzzle. This holds true whether the mortar be dug-in or mounted in an armoured vehicle. This must be regarded as a significant disadvantage with the trend towards greater use of airburst. Breech loading could help to solve the problem, but the implications of such a design has drawbacks which have already been mentioned.

FREE FLIGHT ROCKETS

General Characteristics

The use of rockets to provide artillery support has attracted renewed interest recently. The revival really began during World War II. Among the rocket systems fielded were the Russian "Stalin organ" and the German "Nebelwerfer". Both were extremely useful even though, by modern standards, they had neither a high degree of system accuracy nor a great range capability. Since that time the Soviet Bloc countries have pursued their faith in rockets while Western nations have been somewhat slower to recognise what has proven to be a definite trend.

A rocket motor, in its simplest form, is a tube open at one end in which fuel is burned. The gases from the burning fuel escape out of the open end: the momentum of the escaping gases causing an equal and opposite reaction on the closed end. There is no requirement for any form of ordnance to provide a chamber for a burning charge and to withstand high firing pressures; however, a launcher is needed to carry the rocket and provide the means of pointing it in the right direction. As the name implies, a Free Flight Rocket (FFR) is not a guided weapon, the rockets following a normal ballistic trajectory after launch.

Rockets have certain advantages and disadvantages in terms of rate of fire. On one hand the ability to saturate an area with fire for a short period of time exceeds that of a gun. As an example eighteen Soviet BM21 122 mm Multiple Launch Rocket Systems (MLRS) can deliver more than 16 tonnes of HE onto a target in 20 seconds. Even 155 mm gun systems with a burst fire capability cannot match this without concentrating 4-5 times the number of rocket systems. On the other hand sustained rates can prove difficult with rocket systems because of the reload times. For example the BM21 takes about ten minutes to reload. Although the launch vehicle for a rocket system can be much lighter than an SP gun, for reasons already given, the ammunition itself is much heavier and bulkier. There are two reasons for this. Firstly, for a given range, more propellent is needed than for a gun; secondly, the rocket motor travels with the rocket after launch. The lower accelerations on rockets compared with gun ammunition allows for higher capacity warheads with pre-formed fragments giving rockets an advantage in lethality. Rockets also have greater potential as carriers compared with guns and mortars which is particularly important with the increasing emphasis on sub-munitions.

Compared with guns, rockets are generally inferior in terms of accuracy and consistency, as well as range coverage. The accuracy and consistency of rockets is affected by the wind and thrust misalignment and the difficulty of producing rocket motors to produce fire as consistent as that from guns. The relatively poor range coverage of rockets is a function of their flat trajectory, high muzzle velocity and fixed charge. To an extent this disadvantage can be overcome by the use of spoilers to alter the air flow over the rocket and change the shape of its trajectory.

SUMMARY

For reasons of logistics, training and costs of development it is always desirable to restrict the number of different types of indirect fire weapon systems to a minimum. The problem is that the fewer the number the more versatile the systems retained must be. It also follows that the more versatile the system the more likely it is to be an inadequate compromise for specific tasks. To illustrate this point the tables at the Appendix to this chapter summarises the advantages and disadvantages of guns, mortars and free-flight rockets. It is difficult to exclude completely any of the different types of indirect fire weapon systems, although the balance will obviously vary from army to army depending on their individual needs. Guns will be covered in more detail in Chapters 4 and 5; mortars in Chapter 6 and rockets in Chapter 7.

APPENDIX 1 TO CHAPTER 3

SUMMARY OF ADVANTAGES AND DISADVANTAGES OF DELIVERY SYSTEMS

SYSTEM	ADVANTAGES	DISADVANTAGES
M O R T A R S	High level of mobility. Very good lethality against un-protected troops. Simple, inexpensive, easy to operate and maintain. Silent flight and steep angle of descent good for demoralisation. Ideal for delivering chemical pay-loads. Very good rate of fire. Weapon weight/bomb weight ratio very good. Quick into and out of action. Easy to dig-in and conceal.	No low angle or direct fire capability. Comparatively easy to locate. Limited range capability. Shape of bomb not suited to carrier tasks. Shape of bomb not suited to hollow charge warhead. High angle fire results in long time of flight. Requirement to "bed-in". Breech loading system needed if weapon is to be mounted under armour. Safety in loading.
G U N S	Good range coverage. Robust reliable and well-proven. Good accuracy and consistency. Nuclear capability possible (not below 155 mm). More difficult to dig-in and conceal than mortars. Good response times. Good rate of fire, especially for sustained periods. Direct fire capability.	Limited capability against hard targets with conventional HE projectiles. Protection only possible at the expense of mobility. Weight penalty for increase in range with conventional projec-tiles. Greater logistic load than mortars. Poor airportability if SP.

SYSTEM	ADVANTAGES	DISADVANTAGES
R O C K E T S	Increase in range incurs a smaller weight penalty than with guns. Nuclear capability. Long range. Lethal warhead. Greater potential for sub-munitions than other systems. Large weight of fire in a short period.	Accuracy and consistency not as good as guns. Poor range coverage. Heavy ammunition increases logistic load. Poor sustained rates of fire. Response time not as good as a gun or a mortar.

SELF TEST QUESTIONS

QUESTION 1   Can nuclear projectiles be fired from gun systems?

             Answer   ....

QUESTION 2   List the advantages of towed guns compared with SP guns.

             Answer   ....

                       ....

                       ....

                       ....

QUESTION 3   Why is fin stabilisation preferred to spin stabilisation in
             most mortar systems?

             Answer   ....

                       ....

                       ....

                       ....

                       ....

                       ....

QUESTION 4   What is the main disadvantage in increasing the range of a
             mortar by rocket assistance?

             Answer   ....

                       ....

                       ....

QUESTION 5   Why are supersonic muzzle velocities avoided in the design
             of orthodox mortar systems?

             Answer   ....

                       ....

                       ....

                       ....

QUESTION 6   Why is mobility an important feature of mortar systems?

Answer   ....

....

QUESTION 7   Compare the suitability of guns and rockets for:

a. Close support tasks.

b. Depth support tasks.

Answer   ....

....

....

....

....

QUESTION 8   What is the difference between a mortar and a gun?

Answer   ....

....

....

ANSWERS ON PAGE 187

4.

Ordnance

INTRODUCTION

The external appearance of a gun and the components from which it is made will differ depending on the type of weapon and its purpose. Despite these differences all guns are constructed along similar lines. The two main parts or groups of components in a gun are the carriage or mounting and the ordnance. The carriage or mounting supports the ordnance, provides stability for the gun on firing, includes the arrangements for pointing the gun in the required direction and in some cases a means of transporting the ordnance. Carriages and mounting will be covered in detail in Chapter 5. The ordnance provides a vessel to contain the force of the burning charge in such a way that the energy produced is transmitted safely and predictably to the projectile. It also includes the means of imparting direction and stability to the projectile. The purpose of this chapter is to describe the components of the ordnance and to discuss the important features of its design. The main components of the ordnance are the barrel and its attachments, the breech, and the firing mechanism.

THE BARREL

Rifling

A barrel is essentially a tube of steel through which the projectile passes on firing. The inner surface of the barrel is called the bore and normally it is grooved or "rifled". Smooth bores do exist, but their application is mainly reserved for tank guns, especially those firing fin-stabilised projectiles. The rifling in the bore is engraved along a helical line and may have a constant or a variable twist.

A bore with a constant rifling twist is characterised by the constant slope of the grooves in relation to the axis of the bore, which is the imaginary line along the centre of bore throughout its length. Progressive rifling has a variable relationship between the slope of the grooves and the axis of the bore with the slope increasing towards the muzzle. Progressive rifling can be used to reduce the

pressure on the lands from the driving band around the point of maximum gas
pressure in the barrel yet retaining the ability to impart sufficient spin to the projectile before it leaves the bore. The theoretical advantage is that a shorter barrel can be used without degrading the stability of the projectile in flight. As we
shall see, however, there are other conflicting requirements with regard to how
and where the different levels of pressure are achieved.

The raised surfaces of the rifling are called the "lands" and the diameter of the
bore, excluding the depth of the lands, is the measurement that give the barrel,
and indeed the weapon, its calibre. The purpose of the rifling is to impart spin
to the projectile as it travels along the bore. The projectile is fitted with a driving band made of softer material than the rifling. As the projectile moves forward the driving band is engraved by the rifling, the band itself receiving grooves
corresponding in cross section to the lands. The engraving on the driving band is
then forced to follow the twisting path of the rifling as it moves up the bore with
the result that the projectile acquires spin. Two incompatible requirements must
be resolved in deciding on the depth of the rifling. On one hand it is advisable to
have deep grooves to improve the guidance of the projectile through the bore and
reduce the sensitivity of the rifling to wear; conversely, shallower grooves make
it easier for the driving band to engage in the rifling and the engraving left on the
driving band once the projectile leaves the bore produces less air resistance in
flight.

Sequence of Events on Firing

The sequence of events that take place inside a gun on firing is the subject of that
part of the study of the motion of projectiles called Internal Ballistics. Internal
Ballistics also includes the study of the ballistic properties of propellents and
this aspect is covered in Volume III.

When a gun fires the burning charge produces gases at very high pressures to
accelerate the projectile through the bore until it leaves the muzzle of the gun at
a predetermined muzzle velocity. This muzzle velocity is the basis of the even-
tual range that a projectile of given mass and shape will eventually achieve. In
designing a gun, calculations are made to ensure that the muzzle velocity required
is produced in the most regular, efficient manner. The energy produced by the
charge is expended in performing several different types of work. Much of the
energy is used in imparting velocity to the projectile. It is also expended on the
work done in rotating the projectile, overcoming the friction of the driving band
against the wall, displacing the mass of the charge and gases, and displacing the
mass of the recoiling parts of the gun. Some of the energy is also lost as heat into
the barrel, breech, the projectile, as well as the cartridge case if one is used.

At the instant that the charge is ignited the propellent begins to burn in the confined space the forward and rearwards limits of which are defined by the driving
band of the projectile and whatever method of obturation is used to prevent the
rearward escape of gases. The rate of burning increases in proportion to the
rate of increase in pressure until the gas reaches "shot start pressure". Shot
start pressure is the pressure at which the projectile is moved forward. As the
projectile moves down the barrel the space available for the gases increases thus

reducing the rate of increase in pressure. The point of maximum pressure is reached when the pressure loss caused by the space increase is equal to the pressure increase from the burning propellent. Thereafter the pressure in the bore begins to drop. Meanwhile the projectile continues to accelerate and continues to accelerate even after the charge is all burnt; however, the rate of acceleration decreases until retardation occurs just outside the muzzle. Figure 17 shows the relationship between pressure, distance travelled by the projectile and the velocity of the projectile.

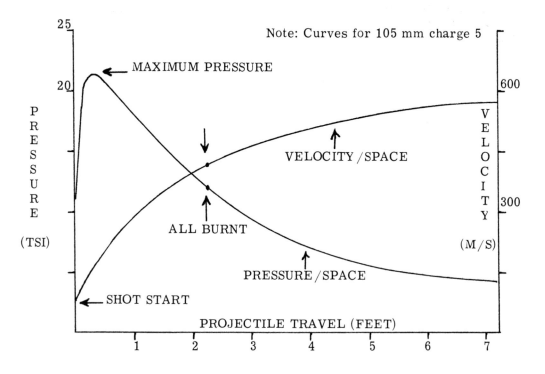

Fig. 17.   Pressure/velocity/space curve

During the period of projectile travel in the bore about 25-35% of the energy produced by the charge is consumed. The remainder is discharged into the atmosphere after the projectile leaves the muzzle. It can be seen from Fig. 17 that it may be possible to increase the muzzle velocity of the projectile by lengthening the barrel and thus increasing the period of time the projectile is influenced by the gas pressure. This method of improving muzzle velocity does have disadvantages because beyond certain limits the increase gained is not justified when balanced against the disadvantages of having a longer barrel. Figure 17 also illustrates this point, for it is apparent that velocity increases are more gradual as the projectile moves away from the all burnt position.

The all burnt position itself is a most important consideration in the design of gun systems. If the all burnt position is too far forward in the barrel the likelihood of a distinctive muzzle flash increases, thus raising the possibility of

detection. If the all burnt position is outside the muzzle there is a danger that
the breech could be opened before all the propellent has been consumed. This
possibility needs close attention in the design of guns and their charge systems,
especially for guns where the breech opens automatically after firing. There are
other reasons for keeping the all burnt position well back in the bore, not the
least of which is the improvement gained in round-to-round muzzle velocity variations. Clearly the all burnt position will also affect the location and magnitude of
the stresses on the barrel. When one considers that, even in a 105 mm field gun
the pressures needed to unleash a projectile from the muzzle at several hundred
metres per second can be well in excess of 20 tons/in^2 , it is understandable
that the question of stresses on barrels receives great attention.

BARREL CHARACTERISTICS

Barrel Life

The main characteristics required of a gun barrel are that it must have long life,
strength, stiffness, be of a suitable mass and have an appropriate centre of
gravity. The service life needs to be as long as practicable which means that it
should be able to fire a high number of rounds before wear renders it incapable
of firing properly stabilised projectiles at required muzzle velocity. There are,
however, some conflicting considerations in the need for long life. The sheer
cost of extending the life of a barrel by the application of special barrel cooling
techniques, cool burning propellents and expensive bore surface finish materials
may simply be uneconomical. The overriding considerations must be the operational requirements, in particular the rates of fire needed and the assessment
of how often it is convenient to change barrels. Nevertheless these operational
requirements still have to be balanced against cost and ease of manufacture.

Strength

It goes without saying that the strength of the barrel must be such that it can
withstand firing stresses without failing under operational conditions. The degree
of strength required can be achieved in a number of ways such as the use of
alloying additions to the steel, by pre-stressing the barrel or by virtue of the
barrel design itself. Normally a combination of all three are used. The required barrel mass is mainly det ermined by strength considerations but stiffness
and the overall stability of the equipment on firing are also considerations.

Stiffness

A barrel must have adequate stiffness, or 'girder strength' as it is commonly
called, so that it does not bend under its own weigh t. This characteristic is particularly important for guns with comparatively long barrels. For a given grade
of steel the girder strength requirement is usually met by the selection of a barrel contour that reduces the barrel thickness from the breech to the muzzle.

Centre of Gravity

Rear trunnions are preferred in most modern guns. The further the barrel's
centre of gravity is from the trunnions the greater the out of balance moment
caused by the barrel and the breech not being pivoted at its point of balance.
Ideally, therefore, the centre of gravity should be as close to the trunnions as
possible. Balancing gears are normally used to solve this problem in modern
equipments and these will be covered in more detail later in Chapter 5. Another
technique is to increase the mass of the breech to provide a counterweight. The
disadvantage of this approach is that it increases the weight of the equipment;
however, the extra weight can be advantageous because it improves the stability
of the gun. The subject of gun stability is discussed in Chapter 5. The tendency
of the barrel to be muzzle heavy can be offset by the mass distribution along its
entire length; however other factors, including those mentioned in earlier para-
graphs, are generally considered more important in deciding on the barrel contour.

BARREL CONSTRUCTION

Wire Wound

Wire wound barrels are now obsolete; however, the techniques used are of interest. Barrels constructed in this manner were made by winding wire tightly
around an inner tube. The wire covering can be extended as far forward from the
breech as is necessary to withstand firing stresses. The method has some advantages, the main ones being that the strength of the wire used can be carefully
controlled during manufacture. Furthermore any failure in the wire tends to be
localised, unlike a crack in solid steel. The disadvantages are that usually some
form of outer tube or jacket must be shrunk onto the barrel to give it girder
strength and the whole process is expensive and slow. Nevertheless the main
reason for this method becoming obsolete is that there are now better techniques
available for pre-stressing barrels.

Built-Up Barrels

Another approach to barrel construction is to build-up the barrel using two or
more tubes slipped one over the other. Before assembly the outer tube has a
slightly smaller diameter than the inner tube. The outer tube is heated and a
cold inner tube is then fitted into it. On cooling, the surfaces of the two cylinders
are compressed together. The result is that between the tubes there is pressure
produced caused by tangential compressive stresses in the inner tube and tensile
stresses in the outer tube. The overall effect is that the compressive force in
the bore layers enables the barrel to withstand higher firing stresses than would
be possible if the barrel were constructed from a single metal tube of equivalent
thickness. This only holds true if the single tube cannot be pre-stressed by some
other process, such as autofrettage which will be explained later in this chapter.
In the past the built-up barrel method had some advantages, especially in the
manufacture of very large barrels for which other methods of strengthening presented insurmountable manufacturing difficulties.

Loose Barrel/Loose Liner

Another method of augmenting barrel strength is to fit a jacket over a part of the barrel to provide additional strength to the highly stressed parts of the barrel. The jacket also provides a form of longitudinal support. The term loose barrel is a little misleading because the barrel and the jacket are a close fit and the two are firmly clamped together to prevent the barrel rotating as the projectile moves along the bore. The advantage is that the barrel can be removed comparatively easily for replacement or in the case of pack guns to facilitate stripping into transportable loads. The loose liner concept is an earlier version of the loose barrel. The difference is that loose liners have jackets extending along the full length of the barrel. There is, of course, a weight penalty when compared to the loose barrel concept. The QF 18/25 pounder had a loose liner whereas some versions of its successor, the 25 pounder, had loose barrels.

Composite Barrels

A composite barrel is not unlike a loose barrel except that individual segments of the barrel are fabricated from different grades of steel; the grades depending on the gas pressures that have to be withstood. Composite barrels can be made with rifled bores or with smooth bore muzzle extensions to increase the muzzle velocity. The idea, behind what seems like a complicated concept, is that sections of the barrel can be interchangeable as dictated by wear in the bore. The disadvantages are its complexity, the problems in ensuring a tight gas seal between the segments and the difficulties in achieving precise coincidence of the lands and the grooves between adjacent segments. Nevertheless, composite barrels have been produced successfully for anti-aircraft and anti-tank artillery. Composite barrels, like all of the barrel construction techniques mentioned so far, have generally been replaced in favour of the monobloc form of barrel construction.

Monobloc

Monobloc barrels are made from a single forging without any jacket or liner. It is a commonly used technique nowadays because of the ease of manufacture and the advances made in metallurgy in creating alloy steels with reliable, high levels of resistance to gas pressures in the bore. Modern techniques for producing monobloc barrels are remarkably quick. Rotary forge machines exist that are capable of both hot and cold forging of hollow or solid cylinders into barrels of lengths in excess of 10 metres. The complete process is less dependent on operator skill than it used to be because of the application of numerical control systems programmed to accomplish the complete operation. Forging process times in the order of ten minutes for a 105 mm barrel are possible. In the cold forging applications, rifling can also be included as part of the barrel forming process. This can be achieved by forming the barrel over a mandrel device containing a mirror image of the desired depth and helical twist of rifling.

For a given grade of steel, to achieve greater strength in a monobloc barrel without fire-stressing, it is necessary to manufacture it with thicker walls. Besides making the barrel heavier and more expensive there are limits to the

usefulness of this approach. The outer layers of steel in a barrel tend to be
under-stressed compared to the inner layers. Beyond a certain ratio of overall
barrel diameter to bore diameter there is no significant increase in strength.
Thus one of the basic drawbacks of monobloc barrels is that because of the non-
uniform participation of the barrel wall in resisting pressure there comes a
point where limiting pressures are acting on the inner surface when the outer
layers are not contributing. The alternatives are to use higher tensile steel, to
pre-stress the barrel, or both. The accepted method of pre-stressing barrels
nowadays is by the process of autofrettage.

AUTOFRETTAGE

Stresses on Barrels

Five different types of stresses act on gun barrels. Girder stress, radial stress,
circumferential stress, longitudinal stress and torsional stress. Girder stress
is a bending stress placed on a barrel by virtue of its length and weight. A barrel must, therefore, be stiff enough to prevent bending under its own weight. On
firing, gas pressure in the bore imposes a radial stress outwards on the walls
of the barrel. Gas pressure also produces circumferential stress which acts
tangentially to the circumference of the bore at any given point. As the projectile moves up the bore two other stresses are produced. The first is longitudinal
stress which is caused by the forward movement of the driving band and the difference in pressure between the front and the rear of the band. Its effect is to
stretch the barrel longitudinally; however, the effect is very much localised and
follows the progress of the projectile. The second type of stress associated with
the movement of the projectile is torsional stress. Torsional stress is generated by the rotation of the projectile as it moves up the bore. It produces a twisting effect in the opposite direction to the twist of rifling.

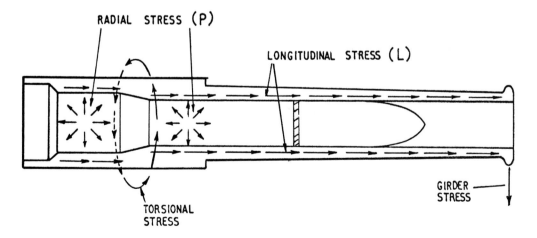

Fig. 18.   Stresses in gun barrels

It is difficult to determine exactly who invented the autofrettage technique. The strongest claims are for the French in the early 1900s, although other French claims that are more difficult to substantiate go back as far as the 1860s. Obviously, the word "autofrettage" is of French origin, its English translation being "self-hooping". Whatever the doubts about the date of origin of the technique, the first autofrettaged gun was certainly a French 140 mm produced in 1913.

In outline the autofrettage process is as follows. The start point is a single steel tube of internal diameter slightly less than the desired calibre. The tube is subjected to internal pressure of sufficient magnitude to enlarge the bore and in the process the inner layers of metal are stretched beyond their elastic limit. This means that the inner layers have been stretched to a point where the steel is no longer able to return to its original shape once the internal pressure in the bore has been removed. Although the outer layers of the tube are also stretched the degree of internal pressure applied during the process is such that they are not stretched beyond their elastic limit. The reason why this is possible is that the stress distribution through the walls of the tube is non-uniform. Its maximum value occurs in the metal adjacent to the source of pressure, decreasing markedly towards the outer layers of the tube. The strain (or the change in dimensions) is proportional to the stress applied within the elastic limit; therefore the expansion at the outer layers is less than at the bore. Because the outer layers remain elastic they attempt to return to their original shape; however, they are prevented from doing so completely by the now permanently stretched inner layers. The effect is that the inner layers of metal are put under compression by the outer layers in much the same way as though an outer layer of metal had been shrunk on. The next step is to subject the strained inner layers to low temperature heat treatment which results in the elastic limit being raised to at least the autofrettage pressure employed in the first stage of the process. Finally the elasticity of the barrel can be tested by applying internal pressure once more, but this time care is taken to ensure that the inner layers are not stretched beyond their new elastic limit.

One advantage of the autofrettage process is that for a given maximum firing pressure a cheaper, lower grade of steel can be used in the barrel. Another advantage is that for a given grade of steel the thickness of the barrel can be reduced with a consequent saving in weight and cost. For example, a barrel with an overall external diameter 50% greater than the diameter of the bore can be strengthened by autofrettage to a level that would require an increase in wall thickness approaching 50% in a non-autofrettaged monobloc barrel. A further advantage is that the effect of compression on the inner layers of the barrel tends to close any tiny cracks in the surface of the bore, thus reducing the possibility of fatigue failure during the life of the barrel. For lighter gun barrels the increase in fatigue life can be in excess of 200%. There are several different approaches to the autofrettage process. They differ primarily in the manner in which the internal pressure is applied to the bore. These approaches can be divided into two groups: Hydraulic Autofrettage and Swage Autofrettage.

Hydraulic Autofrettage

Hydraulic Autofrettage is effected by introducing high pressure fluid into the bore

to achieve the stresses required.   The fluid commonly used is a mixture of
glycerine and water because of its stability at high pressures.   Different countries use different rigs for hydraulic autofrettage and an example of a type of
rig used by the United States is in Fig. 19.

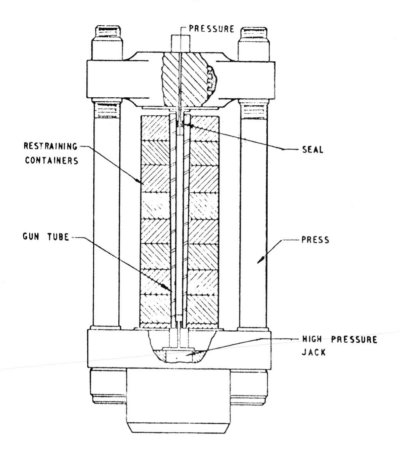

Fig. 19.   Rig for hydraulic autofrettage

Some high strength steels have yield points approaching 70 tons f /in^2, this poses
problems in achieving the necessary hydraulic pressures.   Moreover, even if
the pressures can be produced there are considerable problems in sealing the
rig.   Another inherent disadvantage of hydraulic autofrettage is that different
wall thicknesses along the length of the barrel will result in variable levels of
pre-stressing.   As explained earlier in this chapter, a constant wall thickness
along the length of the barrel is not needed: therefore, if weight is a problem
there may be an added requirement to reduce the thickness of the barrel towards
the muzzle during the finishing process following autofrettage.   The Swage
Autofrettage technique is a means of overcoming some of these difficulties.

Swage Autofrettage

In the swage method an oversized swage or mandrel is forced through the bore by
a hydraulically-operated ram.   The pressures needed to overstrain the inner
layers of the barrel are a function of the grade of steel used in the barrel, its
thickness, the difference between the initial size of the bore and the swage; and
the contact area between the bore and the swage.   For a given level of pre-
stressing the hydraulic pressures needed to push the swage through the bore can
be more easily achieved than the pressure levels needed for Hydraulic Autofrettage.   Furthermore, the swage method presents the designer with the option
of only autofrettaging the length of barrel that requires pre-stressing.   The diagram at Fig. 20 illustrates the swage technique.

mandrel   hydraulic press ram

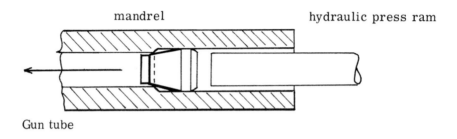

Gun tube

Fig. 20.   Swage autofrettage

BARREL WEAR

Despite the advances made in barrel manufacturing techniques the problem of
barrel wear remains.   Wear is caused by the chemical action of the hot, high
pressure gases from the propellent in the bore and the abrasive action of the
driving band as it passes through the barrel.   The results of wear is a decrease
in the initial resistance to shot start pressure resulting in a lower maximum
pressure and consequently a decrease in muzzle velocity.   The difference in
muzzle velocity occasioned by wear can be calculated and corrections made for
it either on the gun itself or during the production of gun data for the engagement
of targets.   When the wear in the barrel becomes extreme, unacceptable inconsistencies in muzzle velocity are likely to occur.   Additionally, if the rifling is
badly worn the driving band is likely to fail and the projectile may not be adequately stabilised.   If the projectile is not stabilised along its trajectory the air
resistance increases and the projectile will fall short of its target.   The effect is
not predictable and cannot, therefore, be corrected in the same way as a loss of
muzzle velocity.

The wear caused by the action of hot, high pressure gases in the bore is called
erosive wear.   Often the effects of erosive wear are localised: in which case it

is called erosive scoring. Erosive scoring appears and develops quickly at any point of imperfection in the barrel. It can also be induced at the location of any failure in the seal between the driving band and the rifling. A further classification of erosive wear is annular erosion, which appears as a localised, circular enlargement of the barrel at the point where the forward edge of the cartridge case is positioned in the chamber of QF equipments. The type of wear resulting from the friction between the projectile and the bore is called abrasive wear. Abrasive wear will gradually remove the metal from the surface of the bore, with the mechanical attrition on the driving side of the lands resulting in the rounding of the lands. The prevention of abrasive wear is achieved by ammunition design resulting in less friction in the bore; however, erosive wear is usually the most significant form of wear, especially in guns producing long ranges and high rates of fire.

The means of combatting this problem involve a choice of one or more of the following options: wear additives to form an isolating thermal layer on the surface of the bore; the use of cooler propellants; the use of steel with a greater resistance to erosion; and barrel cooling techniques. The main problem with the use of wear additives is that, ideally, they should be renewed with every round, although this is by no means insurmountable. A suitable additive such as a mixture of magnesium silicate and wax can be introduced into the chamber behind the projectile. On firing, the lining of the barrel becomes coated with the additive which protects the bore from the erosive effect of hot gases. The results so far in the use of wear additives are most encouraging, but further research is needed. The use of cooler propellents imposes additional constraints on the designer because of the need for a bigger chamber to accommodate the extra propellent needed to achieve a given maximum pressure. Similarly, the use of barrel materials that have a greater resistance to erosive wear has its disadvantages. Most of these materials, such as molybdenum and chromium, are expensive and their application difficult. Chromium plated barrels have been the subject of much research in recent years; however, the abrasive action of the projectile can cause failure in the plating and to date the technique has been reserved primarily for small calibre weapons.

The only remaining option is to provide some means of cooling the barrel. With the trend towards artillery weapons with higher rates of fire, especially sustained rates, the need to solve the problem of barrel overheating has become much more acute. There are two approaches to the problem: water cooling and air cooling. Water cooling has been used successfully in anti-aircraft and naval gun systems where weight is not a major consideration. Its application in artillery weapons providing close and medium support is questionable; certainly for towed equipments. To be effective the water, or whatever coolant is used, must be kept in close contact with the barrel. The cooling system must also cater for circulation of cool water from some form of reservoir and incorporate some provision for the escape of steam. Water cooling may be a possibility for SP guns where the means of transporting the additional load is integral to the system; however, the penalties in space, weight and added complexity could be unacceptable. Air cooling is an attractive approach to the problem because of its simplicity and the savings in weight compared to water cooling.

The difficulty with air cooling is that the radiating surfaces of a gun barrel are often not sufficient to reduce temperatures quickly enough to offset the heat

input from sustained rates of fire. For example, a 105 mm Light Gun firing at
a rate of 4 rounds per minute for 15 minutes followed immediately by 4 rounds
every 3 minutes for a further 15 minutes could reach a barrel temperature of
around 160°C. Even after a pause in firing of 90 minutes the barrel temperature
will still be at 60-70°C depending on the ambient air temperature. Obviously if
the same rates were fired again the barrel will reach 160°C and beyond much
quicker. Erosive wear increases steadily at bore surface temperatures over
about 675°C and is accelerated above 980°C, although wear additives and the surface finish of the bore may help alleviate the problem. Despite the reduction in
bore temperature achieved by these means the heat generated in a high performance gun is such that the maximum bore temperature reached will be mainly a
function of the total number of rounds fired during the weapons battlefield day.
This is particularly true of modern artillery weapon systems that may be required to fire rates approaching 1000 rounds per day. Whether the rates are
produced in short bursts or not is relatively insignificant.

Attempts to cool the barrel by increasing the radiating surface have been tried.
These methods include the use of fins, rings and ribs. Generally the increase
in radiating surface is insufficient to be given serious consideration in gun design. It may be possible to fit a slotted or finned jacket over the barrel and
force cool air between the two; however, the additional weight and complexity
would be a disadvantage. Any increase in barrel thickness would also help because of the additional heat sink provided. Unfortunately, for artillery systems
providing close and depth support the weight increase precludes this approach.

The consideration of bore temperatures goes beyond the question of wear. Safety
is another factor because it is possible, at high rates of fire, for the bore temperature to cause "cook-off" or spontaneous ignition of the charge and in very
extreme circumstances, detonation of the projectile itself. With the trend towards high rates of fire it is desirable to have some means of monitoring bore
temperatures. The US 155 mm M198 has this facility.

BARREL FATIGUE

The latest generation of artillery weapons are required to fire at greater ranges
than before yet at the same time remain light enough to meet strategic, tactical
and battlefield mobility requirements. To meet this need barrels have been constructed from steels of much higher yield strengths to enable them to cope with
the high bore pressures needed to produce the additional ranges. Yield strengths
in the order of 1200 MPa are possible but usually at the expense of a great reduction in fracture toughness. The result has been an increased tendency for barrels to reach their life condemnation limits before they have to be condemned
because of erosive or abrasive wear. In other words the barrels used now have
a greater propensity to fail because of fatigue or fracture under repeated stresses at stress levels below their tensile strength.

A network of tiny cracks or crazing can be found in most gun bores; however, at
the high bore pressures used nowadays the propagation of crazing into larger
cracks is accelerated. Eventually the fatigue cracking can cause catastrophic
failure of the barrel on firing. Understandably the phenomenon is receiving

increased attention and it is evident that there are two aspects to the problem. The first is to predict the number of times the gun can be fired before barrel fatigue becomes critical. The second is to establish which steels are the best in terms of fracture toughness and their ability to contain the growth rate of fatigue cracking. At present there are difficulties in producing accurate information, the main problem being insufficient evidence of the effect of temperature on fatigue and insufficient evidence of the crack propagation in autofrettaged barrels.

BREECH MECHANISMS

General Description

A breech mechanism is a mechanical device for closing off the chamber in the bore of a gun. It performs the following functions: it withstands the rearward thrust of gas pressure when the gun is fired; and it houses the firing mechanism. In BL guns it also provides the means of obturation, while in QF guns it supports the cartridge case once it is loaded and provides the means of extracting the cartridge case after it has been fired. There are five important characteristics to be satisfied in the design of a breech. It must be reliable and durable under the demands of operational conditions over a wide range of climatic conditions. The design must be safe to operate. In particular the arrangements for opening and closing the breech must be such that it is impossible for the gun to be fired when not fully closed or for the breech to open accidentally on firing. The design must also permit quick and easy loading, unloading and firing. Assuming that these basic requirements are met the design must be as simple as possible. Finally its design should lend itself to production in quantity and easy replacement of the component parts.

Types of Breech Mechanisms

Breech mechanisms can be grouped into two categories, screw mechanisms and sliding block mechanisms. Screw mechanisms are usually fitted to BL systems and sliding block mechanisms are employed with QF equipments. There are many variations on the two basic categories of breech mechanisms. Furthermore, it is possible to design a sliding block mechanism for use in a BL equipments. Some of these variations will be discussed in this chapter.

Screw Mechanisms

The major components of a screw mechanism are shown in Fig. 21. The breech screw is held in a carrier attached to the breech ring on one side to enable it to swing freely during opening and closing. In Fig. 21 the carrier is mounted on the side of the breech ring but if required it can be mounted to give vertical movement during loading drills. On closing, the screw threads on the breech screw engage in corresponding threads on the inner surface of the breech ring. The threaded surface of the breech screw is divided into an even number of sections with alternate sections cut away. The threaded surface inside the breech ring is

similarly designed except that the cutaway sections of the breech ring are opposite the threaded sections of the breech screw.   When the breech is closed by
activating the loading breech mechanism (LBM) see Fig. 21, the breech screw is
then turned through an arc until both sets of threads are fully engaged.   This form
of thread is called an "interrupted" thread or "slotted screw" thread and is the
most commonly used form of thread in modern screw breech mechanisms.   The
pitch of the thread ensures that friction will prevent the breech screw rotating and
opening when the gun is fired.   Some form of gearing is required so that unlock-
ing and swinging the breech screw can be achieved with a single movement of the
LBM.

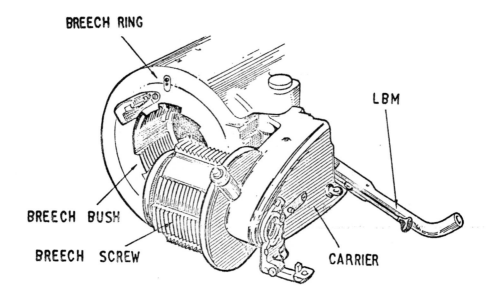

Fig. 21.   Screw breech mechanism

Although not apparent in Fig. 21, the next diagram Fig. 22 shows the arrangements for obturation in a conventional screw mechanism.   In a BL system such
as this the charge is contained in some form of combustible material, normally
a bag.   The obturator is a resiliant pad made of neoprene, usually with a glass
fibre filler.   The pad is in a coned seating and is secured to the breech screw by
the bolt vent axial and the mushroom head.   On firing the mushroom head is
forced to the rear by gas pressure thereby squeezing the pad against the front of
the breech screw.   The pad is forced to expand radially forming a tight seal
against the rearward escape of gases.

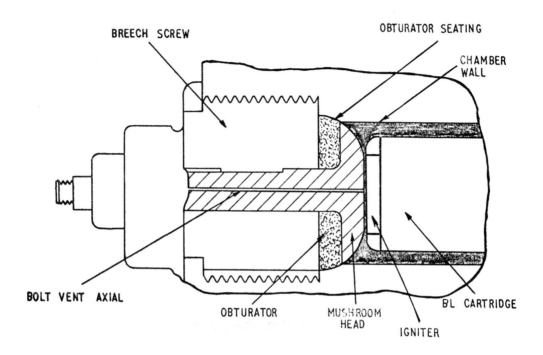

Fig. 22.   BL obturation

An advantage of the screw mechanism is that compared with a block mechanism, for a given performance, it can be considerably lighter. The reason for this is that, on firing, the longitudinal stress from gas pressure is distributed over the whole of the threads. The other main advantage is that, because the method of obturation can be incorporated in the mechanism, there is no requirement for cartridge cases. Nevertheless the concept is not without some disadvantages. Because it is comparatively complex it is more difficult to manufacture. In addition, it is slower to operate and not as easily adaptable for automatic or semi-automatic loading.

Sliding Block Mechanisms

The main components of sliding block mechanisms are: the breech block; the breech ring; the LBM; the extractors; and devices to ensure safety. The breech block can be designed to slide either vertically or horizontally in the breech ring. Fig. 23 shows a vertically sliding block. This type of block is sometimes called a drop block.

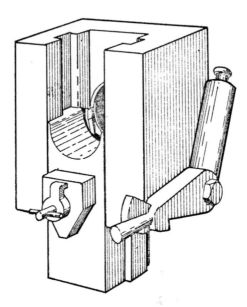

Fig. 23.   Vertical sliding block

The main advantages of sliding block mechanisms are their simplicity and safety together with their ease and speed in operation. They are safer because, on closing, there is less chance of the loader's hand being caught in the breech than with a screw mechanism. A hand left at the entrance to the breech will be swept clear safely by the vertical or horizontal motion of the sliding block. Their simplicity and speed make them readily adaptable to automatic operation.

The disadvantages of conventional sliding block mechanisms are that they require the use of cartridge cases to provide rear obturation and for a given level of rearward thrust on firing they need to be heavier than screw mechanisms. The reasons for this will be explained later in this chapter. Sliding block mechanisms are widely used in equipments producing high rates of fire, particularly in calibres of 105 mm and below where the weight of the cartridge case is not a limiting factor in the rates of fire achieved.

The space available for loading drills throughout the limits of elevation and depression of the gun may dictate whether a vertical or a horizontal sliding block is used. In this regard a vertically sliding block usually offers more scope because it moves in a plane less likely to include obstructions. On the other hand, horizontally sliding blocks require less effort to operate and the effort required remains constant regardless of the attitude of the barrel. The fact that both horizontal and vertical sliding blocks are still being used indicates that no definite trend towards a preferred solution has emerged.

Breech rings for sliding block mechanisms can be of either the "tied jaw" or "open jaw" type. Both are shown in Fig. 24. Tied jaw breech rings are inherently much stronger than open jaw rings because their shape makes it easier to provide the rigidity needed to withstand firing loads. Both types have guide slots that mate with ribs on the sliding block as it opens and closes. The ribs and slots are inclined so that a small amount of forward travel is achieved when the breech is closed. This design feature ensures that the cartridge case is firmly supported. The design can include one or more pairs of guide slots and ribs. In open jaw breech rings their surfaces accept rearward firing loads but in tied jaw rings this is only so if the ribs are in contact with the breech ring on firing.

The greater the number of ribs the greater the number of thrust surfaces to distribute the firing loads over the breech ring. The disadvantage of sliding block mechanisms with multiple thrust surfaces is that it is difficult to machine the surfaces at tolerances to ensure all surfaces accept a part of the load. Regardless of the number of thrust surfaces this type of breech cannot match the efficiency with which the interrupted threads of screw mechanisms can distribute firing loads. Hence for a given load the sliding block mechanism must be heavier.

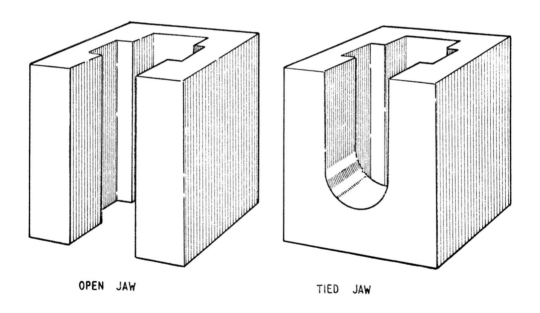

OPEN   JAW   TIED   JAW

Fig. 24.   Open jaw and tied jaw sliding block type breech rings

As mentioned earlier, QF obturation is the method of obturation used in conventional sliding block mechanisms. The rearward escape of gases is prevented by enclosing the charge in a tapered metal case. The case is normally made of brass or steel resiliant enough to expand on firing to form a tight seal against the chamber wall and to contract for easy extraction from the breech after firing. The Swedish Army uses 155 mm plastic cartridge cases with a brass base. Full brass cases are preferred in British Service, the alloy being 30% zinc and 70% copper. In other armies, notably the US Army, steel cartridge cases have been employed successfully. The means of ignition for the cartridge is a primer fitted into the base of the cartridge case. A diagram of the system is shown in Fig. 25.

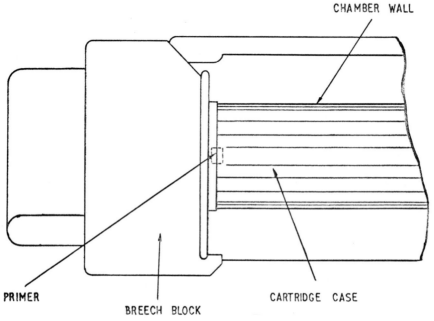

CHAMBER WALL

PRIMER

BREECH BLOCK

CARTRIDGE CASE

Fig. 25.   QF obturation

The demand for increased range with the attendant high chamber pressures to produce it give rise to difficulties in the continued use of metal cartridge cases in the future. The internal pressures that can be withstood by a modern auto-frettaged barrel.can overstrain cartridge cases to the point where they become very difficult to extract. This problem is exacerbated by the use of longer cartridge cases. Extraction problems will seriously degrade the rate of fire and consequently nullify one of the prime advantages of sliding block mechanisms: ease and speed of operation.

One approach to the cartridge case problem is to retain the advantages of the sliding block mechanism and replace the cartridge case with a bag charge. The obturation would then have to be supplied by some other means. A form of metal to metal obturation can be provided by means of an insert in the sliding block designed to make close contact with a metal ring or bush fixed to the chamber-face. This concept has already been put to use in the FH70 155 mm Towed Howitzer.

The extractors fitted to a conventional sliding block breech perform two functions. The first is to unseat the cartridge case after the gun has fired; the second is to eject the spent case. In some equipments they are also used to hold the breech block in the open position. The design of the extractors should satisfy the following requirements:

a. They must allow for smooth, powerful and relatively slow initial movement to unseat the cartridge case.

b. The cartridge case should not be misaligned from the axis of the bore as it is being withdrawn from the chamber.

c. The extractor should be able to unseat and eject the cartridge case without damaging the rim at the base of the case.

d. The final movement of the extractors to eject the case should be fast enough to throw it safely clear of the breech to the required distance.

e. There should be no tendency for the extractors to bounce when the breech is opened.

FIRING MECHANISMS

Firing mechanisms are classified into three groups by virtue of their mode of operation. They are: "percussion"; "electric"; and "percussion and electric". In some armies the term mechanical action is used instead of percussion. American firing mechanisms are called either firing mechanisms or firing locks. Whether it is called a lock or a mechanism is not an indication of the method of obturation of the equipment as is the case with British firing mechanisms. In British Service firing mechanisms used with QF equipments are designated as such, whereas those used with BL equipments are called locks. In the interests of clarity the British understanding of the terms will be used in this chapter. Percussion mechanisms employ the movement of a firing pin to strike the base of a primer to function and initiate the charge. Electric mechanisms initiate the propellent by the use of an electric charge carried by a wire to the primer. Percussion and electric mechanisms as the name implies use a combination of both techniques.

In BL equipments the lock is attached to the rear of the bolt vent axial (see Fig. 22 earlier in this chapter). The means of ignition is a tube positioned in the rear of the bolt vent axial. The lock contains a striker to ignite the tube. On initiation the tube fires pellets of gunpowder through the bolt vent axial to ignite the charge. In QF equipments the firing mechanism is housed in the breech block and contains a firing pin (or an insulated needle in the case of electric mechanisms) to initiate the charge. When the firing lever or lanyard is pulled the pin protrudes through the front face of the breech and strikes the primer housed in the base of the cartridge case (see Fig. 25 earlier in this chapter). All firing mechanisms and locks have some form of safety catch or an arrangement for interrupting the circuit in electric mechanisms. Percussion mechanisms need to be recocked after they have been activated. This can be accomplished by automatic means, by hand cocking, or by the use of a "trip action" mechanism. The advantage of the trip action method is that the mechanism can never be left cocked. When the firing lever is pulled the trip action mechanism drives the striker forward and immediately brings it back to rest in a safe condition. The American M2A2 105 mm Howitzer (QF) has a trip action firing lock. Another American firing mechanism that operates on a similar principle is the strike and hammer type of mechanism. This concept is used in BL equipments such as the American 8 inch Howitzer.

Both electric and percussion methods of operating firing mechanisms are suitable for use. The advantages of an electric mechanism compared with a percussion mechanism are:

a.   It is simpler, lighter and more compact.

b.   It is easier to test for functioning.

c.   Its response to firing is much quicker (0.001 second) compared with 0.50 second for some percussion mechanisms such as trip action.

d.   It is mechanically reliable and less prone to wear.

The disadvantages of electric mechanisms compared with percussion mechanisms are:

a.   Electric contacts must be kept clean.

b.   They are more susceptible to climatic conditions that could introduce moisture or grit into the system.

FUME EXTRACTORS

Fumes from burning propellant can cause problems in SP equipments if they are allowed to flow back into the crew compartment after the breech is opened. Fume extractors are fitted to the barrels of SP guns to ensure that all or most of the fumes are discharged from the muzzle. Fume extractors are cylinders fitted to a segment of the outer surface of the barrel to form a reservoir between the barrel and the cylinder walls. Ports or nozzles are drilled through the barrel to give the gas flow access to the reservoir. The nozzles are inclined towards the muzzle. When the gun is fired the reservoir fills as the gas flow moves along the barrel behind the projectile. The gas flow to the reservoir stops when the pressure in the bore is equal to the pressure in the reservoir.

When the projectile is ejected from the muzzle the pressure in the bore drops rapidly to atmospheric pressure. The result is that the fumes contained in the reservoir are forced back through the inclined nozzles towards the muzzle. The action of the fumes being forced from the reservoir towards the muzzle purges the fumes left in the chamber and in that part of the barrel behind the fume extractor. The size of the fume extractor's reservoir together with the inclination and dimensions of the nozzles must be designed to allow for the purging cycle to

commence before the breech is opened.   Figure 26 shows the action of a typical
fume extractor.

Breech block   Cartridge Case   Reservoir   Projectile

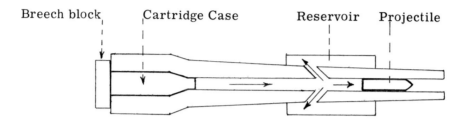

1. Gas drawn into reservoir as projectile passes.

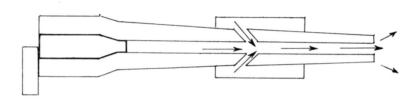

2. Projectile ejected from muzzle and gas drawn from reservoir
    and expelled from muzzle.

Air

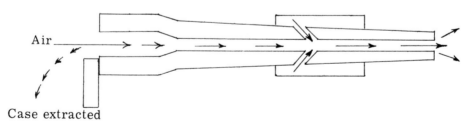

Case extracted

3. Breech opened cartridge case ejected (if used).   Air drawn into
    bore through breech and expelled from muzzle.   Reservoir purged
    of gas.

Fig. 26.   Operation of a fume extractor

MUZZLE BRAKES

General Description

Muzzle brakes first appeared in the mid 19th century as a means of reducing recoil but they fell out of favour with the introduction of recoil mechanisms. They
started to appear again on French guns around the time of World War II. A
muzzle brake is a short cylindrical attachment mounted on the barrel at the
muzzle. It has a centrally bored hole through which the projectile passes and
one or more baffles. Normally, muzzle brakes are screwed onto the barrel with
the screw thread in the opposite direction to the twist of rifling to prevent un-
screwing when the gun is fired. Locking devices are also provided. In some
equipments they are made as an integral part of the barrel.

There are many different designs and methods of construction for muzzle brakes
(see Fig. 27-8). The design can allow for a variable number of slots or baffles
which can be mounted longitudinally or laterally. There are various methods of
construction used producing different levels of efficiency and durability. The
"built-up" type of muzzle brake consists of several plates bolted together. Alternatively the plates can be welded together. "Swaged" muzzle brakes are made
from a single tubular steel forging swaged down to shape and then machined.
Other methods include cast muzzle brakes and muzzle brakes machined from a
solid forging. The latter process, though slow, expensive and restricted to
simple designs, nonetheless produces durable muzzle brakes.

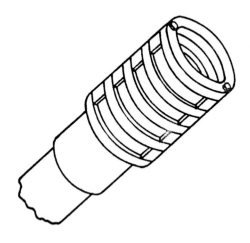

Fig. 27.   Muzzle brake (built-up)

Swaged — single baffle

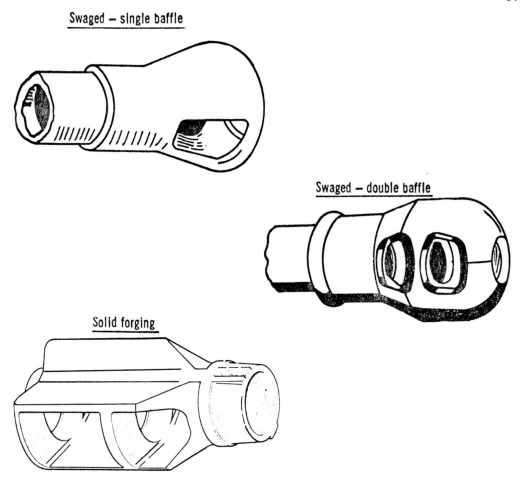

Swaged — double baffle

Solid forging

Fig. 28.   Swaged and forged muzzle brakes

Operation and Uses

When the gun is fired the gases moving behind the projectile strike the muzzle
brake's baffles exerting a force acting in a forward direction.   The effect is to
reduce the amount of recoil energy (see Fig. 29).   For a given gun and ammunition the effectiveness of the muzzle brake depends on the angle through which the
gases are deflected and the size and number of baffles.   An explanation of how a
muzzle brake improves stability will be covered in Chapter 5: however, the use
of muzzle brakes has other advantages too.   First, the recoil mechanism can be
reduced in size because it does not have to absorb as much energy.   Second, it
may be possible to utilise the same carriage for ordnance of larger calibre or to
permit the firing of a higher maximum charge from the same gun.

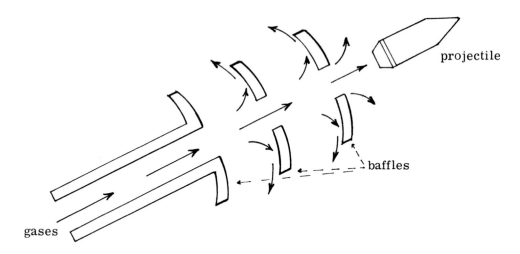

Fig. 29.   Muzzle brake action

Efficiency

As the projectile is about to leave the muzzle of the gun the gases immediately behind it are moving at the same velocity as the projectile.   Once the projectile has
cleared the bore the gases are able to expand.   During this period of expansion
the velocity of the gases increases and, if they are made to deflect from the baffles through an angle, the force exerted will be considerable.   Theoretically, the
greater the angle the greater the force exerted and ideally the gases should be
deflected through 180°.

In practice this is unattainable because of friction and turbulence.   Furthermore
if deflection at 180° occurred the detachment could be endangered.   The efficiency
of the muzzle brake is also enhanced if its diameter is wide enough to enable the
gases to develop their maximum velocity as they expand down to atmospheric
pressure.   This too is impractical because the width of the muzzle brake would
need to be approximately 25 times the diameter of the bore.   Instead, the compromise maximum reached is usually a factor of 5 or less.   A further problem
is that a proportion of the gases follow along behind the projectile and are not deflected by the first set of baffles.   Additional sets of baffles will assist in deflecting more of the gas; however, each baffle only deflects about 60% of the gas that
reaches it.   Eventually a point is reached where additional baffles are not worth
the increased weight and expense.

The definitions of muzzle brake efficiency that are most commonly used are:
gross efficiency, intrinsic efficiency and free recoil efficiency.   "Gross
efficiency" is a measure of the percentage reduction in recoil energy to be absorbed by the recoil system as a result of the muzzle brake fitted.   It does not
take into account the effect of the mass of the muzzle brake itself.   The "intrinsic
efficiency" is the gross efficiency corrected for the effect of the mass of the
muzzle brake.   Its value is less than the gross efficiency figure because part of

the overall reduction in energy is caused by the mass of the recoiling parts (including the muzzle brake). Low efficiency muzzle brakes have intrinsic efficiencies of 20-30% and are comparatively light. High efficiency muzzle brakes have
intrinsic efficiencies of 70-80%. Intrinsic efficiency can be expressed as:

$$I = 1 - \left(\frac{Rb}{Ro} \times \frac{Mb}{Mo} \right)$$

Where: I   = Intrinsic efficiency
         Rb = Recoil energy with muzzle brake.
         Ro = Recoil energy without muzzle brake.
         Mb = Mass of recoiling parts with muzzle brake.
         Mo = Mass of recoiling parts without muzzle brake.

"Free recoil efficiency" is a measure of the percentage reduction in recoil energy
achieved, allowing free recoil during gas action and corrected for the mass of the
muzzle brake. This measure of efficiency takes account of the reduction in recoil energy resulting from the mass of the recoiling parts of the recoil mechanism.
It can be expressed as:

$$Fr = 1 - \left(\frac{Eb}{Eo} \times \frac{Mb}{Mo} \right)$$

Where: Fr = Free recoil efficiency.
         Eb = Free recoil energy with muzzle brake.
         Eo = Free recoil energy without muzzle brake.
         Mb = Mass of recoiling parts with muzzle brake.
         Mo = Mass of recoiling parts without muzzle brake.

Intrinsic efficiency is a more practical figure of merit because free recoil
efficiency is measured by ballistic pendulum and muzzle brake constants that are
independent of the carriage, or mounting. The effects of friction in the carriage
could be high and may well be greater than the sliding friction of the recoiling
mass.

Disadvantages of Muzzle Brakes

Muzzle brakes have some disadvantages. They increase the weight and complexity of the equipment. If retro-fitted it may be necessary to fit some form of
counterweight behind the trunnions to ensure that the gun is correctly balanced.
There may also be a need to make corrections for muzzle velocity, jump and
droop. Gases deflected by a muzzle brake can cause problems with obscuration
at the gun (which is a problem if the gun is being used in a direct fire role) and
in concealing the gun position. A further significant problem is the blast overpressure produced by a muzzle brake.

The blast from a muzzle brake can have a deleterious effect on the gun detachment. The reason for this is that the addition of a muzzle brake increases blast
overpressures to the rear, although there is a reduction in blast forward of the
muzzle. Regardless of the type of muzzle brake used there are definite peaks of

blast intensity in certain directions within a distance of about 30 calibres from the muzzle. Beyond that distance the blast overpressures begin to fall and the blast wave tends to be a sonic wave.

It is in the area behind the gun that the levels of blast overpressures are of greatest interest because that is the area where the gun detachment works. Blast in the vicinity of the breech increases linearly with the increase in aerodynamic index for the muzzle brake being used. The human level of tolerance for blast overpressures is about 2.0 psi and this is the level above which ear drums can rupture. Lung damage can occur at 20-30 psi overpressure. In decibels (db) the levels are 173 and 196-200 respectively. If blast levels are viewed in the extreme, overpressures of about 500 psi can kill outright.

Because damage to the ears appears at the lowest end of the spectrum the acceptable limits of blast overpressures are determined by the levels needed for ear protection. The effect of exposure to blast overpressures even at levels below 2.0 psi can temporarily impair hearing. Prolonged exposure to higher levels can result in a permanent loss of hearing at some frequency levels. The normal means of protection is some form of ear muff or ear plug and these can provide attenuation levels of up to about 35 db. Gun shields can provide a degree of blast protection but they are inadequate for the task with modern high performance towed equipments. Moreover, the additional weight of a gun shield can be a disadvantage, especially for equipments that must be transportable by helicopter. Other means of deflecting blast have been tried with limited success. These include arrangements of baffles mounted at 90° to the axis of the bore either as part of the muzzle brake or further back along the barrel (see Fig. 30).

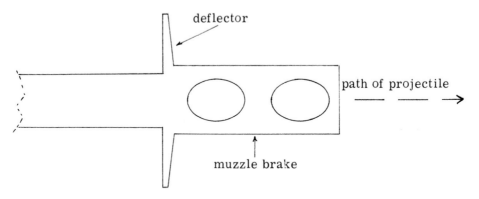

deflector

path of projectile

muzzle brake

Fig. 30.　Deflection of muzzle blast

Recent examples of towed equipments produce blast overpressures that are reason for concern. Moreover, most are fitted with muzzle brakes. To date the problem has not been completely solved. The only alternative seems to be the use of correctly fitted ear protection, accepting the fact that, in operations at least, the remaining risk must be taken. In training the firing of higher charges could be restricted if necessary.

SELF TEST QUESTIONS

Question 1   The carriage or mounting is one of the major groups of components
              in a gun.   What is the other?

              Answer   ....

Question 2   What is "progressive rifling"?

              Answer   ....

              ....

Question 3   What is meant by the term "calibre"?

              Answer   ....

              ....

Question 4   The depth of rifling in a barrel can vary.   What are 2 of the
              considerations in deciding on the depth?

              Answer   ....

              ....

Question 5   Describe briefly the sequence of events that occurs inside a gun
              when it fires.

              Answer   ....

              ....

              ....

Question 6   Approximately how much of the energy produced by the charge is
              consumed while the projectile is still in the bore?

              Answer   ....

Question 7   What are some of the problems that could occur if the "all burnt"
              position is too far forward in the barrel?

              Answer   ....

              ....

              ....

Question 8   List the main characteristics required of a gun barrel.

       Answer   (1)   ....

                 (2)   ....

                 (3)   ....

                 (4)   ....

                 (5)   ....

Question 9   What is "girder strength" and how is it catered for in the design of a barrel?

       Answer   ....

              ....

Question 10   What is the ideal position for the centre of gravity of a barrel?

       Answer   ....

Question 11   List the methods that could be used to solve any tendency of a barrel to be "muzzle heavy".

       Answer   ....

              ....

              ....

Question 12   List 4 methods of barrel construction.

       Answer   (1)   ....

                 (2)   ....

                 (3)   ....

                 (4)   ....

Question 13   List 5 different types of stresses that act on a barrel when the gun is fired.

       Answer   (1)   ....

                 (2)   ....

                 (3)   ....

(4) ....

(5) ....

Question 14 What is "torsional stress"?

Answer   ....

....

Question 15 What is "autofrettage"

Answer   ....

....

Question 16 What is the main advantage of the "autofrettage" process?

Answer   ....

Question 17 List the 2 main approaches to the autofrettage process.

Answer   (1) ....

(2) ....

Question 18 Explain the term "erosive wear".

Answer   ....

....

Question 19 What causes "abrasive wear" in gun barrels?

Answer   ....

....

Question 20 It is possible to water-cool gun barrels but what are the problems involved?

Answer   ....

....

....

Question 21 What is the main limitation in relying on air cooling for gun barrels?

Answer   ....

Question 22 Explain the term "cook-off".

Answer   ....

....

Question 23 State the 2 main considerations in solving the problem of fatigue
in gun barrels.

Answer   (1) ....

(2) ....

Question 24 List the main functions of the breech.

Answer   ....

....

....

....

Question 25 Breech mechanisms can be grouped into 2 categories.   What are
they ?

Answer   (1) ....

(2) ....

Question 26 What is an LBM ?

Answer   ....

Question 27 Explain the term obturation and the manner in which it is achieved
in BL and QF equipments.

Answer   ....

....

....

....

Question 28 List the advantages and disadvantages of a screw breech mechanism
compared with a sliding block mechanism.

Answer   ....

....

....

....

....

....

Question 29 List the advantages and disadvantages of "multiple thrust surfaces"
in a sliding block breech mechanism.

Answer   ....

....

....

Question 30 List 3 of the design requirements for "extractors" in a sliding
block mechanism.

Answer (1) ....

(2) ....

(3) ....

Question 31 What are the 3 main classifications of firing mechanisms?

Answer (1) ....

(2) ....

(3) ....

Question 32 Electric firing mechanisms have some advantages over percussion
mechanisms. List 3 advantages.

Answer (1) ....

(2) ....

(3) ....

Question 33 What is the purpose of a fume extractor and how does it work?

Answer   . . . .

. . . .

. . . .

Question 34 What is the purpose of a muzzle brake and how does it work?

Answer   . . . .

. . . .

. . . .

Question 35 Why is the effectiveness achieved by increasing the number of baffles on a muzzle brake limited?

Answer   . . . .

. . . .

Question 36 What is the approximate theoretical diameter of a muzzle brake for 100% efficiency?

Answer   . . . .

Question 37 Muzzle brake efficiency can be expressed in several ways. List 2 of them and explain what they mean.

Answer   (1) . . . .

(2) . . . .

Question 38 What are the problems that could arise if a muzzle brake is retro-fitted?

Answer   . . . .

. . . .

Question 39 List the main disadvantages of muzzle brakes.

Answer   . . . .

. . . .

. . . .

. . . .

Question 40   Blast overpressures produced on firing can be a safety hazard to
gun detachments and the dangers can be increased by the use of a
muzzle brake.   What are the critical levels of blast for human
safety?

Answer   ....

....

....

Question 41   State 3 methods that can be used to protect gun detachments from
the effects of blast when a gun is fired.

Answer   (1)   ....

(2)   ....

(3)   ....

ANSWERS ON PAGE 188

5.

Carriages and Mountings

INTRODUCTION

A carriage or mounting is the combination of assemblies that supports the ordnance, provides stability for the gun on firing and includes the arrangements for pointing the gun in the required direction. In the case of a mobile equipment it may also provide the means of transportation.

The difference between a carriage and a mounting is that a carriage fires with its wheels in contact with the ground, whereas a mounting does not fire with wheels in contact with the ground. Mountings can be further classified as mobile mountings or self-propelled (SP) mountings. A mobile mounting travels on its wheels but the wheels are either raised or removed before firing and either a plate and/ or girders used to support the weapon (see Fig. 31). A self-propelled (SP) mounting is one which is mounted on a tracked or wheeled chassis with its own motive power. Tracked chassis are usually preferred nowadays; however, a recent example of a wheeled SP gun is shown in Fig. 32.

Two other classifications of mountings should be mentioned. They are: static mountings which are mounted permanently, not on wheels and are normally never moved; and semi-static mountings which, although capable of being moved require specially prepared sites. These types of artillery mounting were mainly used for coastal, garrison, railway and early air defence artillery systems. They are now obsolete and will not be discussed any further.

The main parts of a carriage or mounting are its "superstructure" and its "basic structure". The components of the superstructure are: the saddle (sometimes referred to as the top carriage); the elevating, traversing, and balancing gears (sometimes called equilibrators); the cradle; the recoil system; and the sights. The superstructure supports the ordnance and provides the means of pointing the barrel in the required direction.

The main parts of the basic structure are: the saddle support (sometimes called a bottom carriage); the trails and the articulation system if it is a split trail weapon;

platforms and spades; the wheels, axles, suspension and brakes. Not all of these components are included in every carriage and mounting.

Fig. 31.   Mobile mounting (8 inch)

Fig. 32.   Wheeled SP Gun

SECTION 1.   THE SUPERSTRUCTURE

GENERAL CONFIGURATION

A typical configuration for the superstructure of a conventional gun is shown in Fig. 33. Its function can be explained as follows. The cradle supports the barrel and houses the recoil mechanism, the recoil mechanism being attached to the barrel. It has trunnions which provide the axis about which the cradle and the barrel rotate in the vertical plane when the gun is elevated or depressed. The trunnions fit into bearings in the saddle so that the saddle can support the cradle. The saddle is provided with capsquares to retain the trunnions in their bearings. The saddle itself rests on the saddle support which is part of the basic structure. When the gun is traversed the saddle rotates about its pivot which is held by the saddle support. The balancing gear is interposed between the cradle and the saddle, or sometimes between the cradle and the basic structure. When the gun is elevated or depressed the balancing gear corrects any out-of-balance moment.

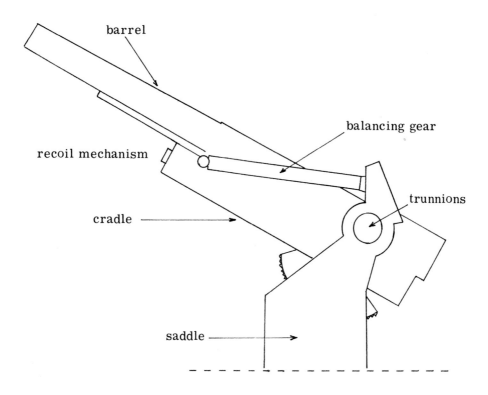

Fig. 33.   Superstructure of a conventional gun

THE CRADLE

The most common type of cradle is the "trough cradle". A trough cradle is a U-shaped box mounted beneath the ordnance with the recoil mechanism positioned within the trough. The cradle in Fig. 33 on the previous page is a trough cradle the upper surfaces of the trough have guideways along their length for the recoiling parts. The front end of the trough is blocked off by a removable cap called the cradle cap and the recoil mechanism piston rods are attached to it. An advantage of a trough cradle is that it affords some protection to the recoil mechanism from damage by shell splinters or small arms fire.

The other type of cradle is the "ring cradle" (also referred to as a yoke cradle). A ring cradle is normally a casting in the form of a cylinder around a segment of the barrel. The barrel and the ring cradle are in contact during recoil and runout. Usually the inner surface of the ring is fitted with sleeves or pads for the barrel to slide against, and the cradle design should allow for their replacement and lubrication. The cradle is built up to house the recoil mechanism externally and the recoil mechanism piston rods are connected to a flange on the barrel. Compared with the trough cradle, the ring cradle is more compact, has a stiffer cross section and is easier to manufacture. Besides the lack of protection for the recoil mechanism, its disadvantages include the existence of a critical clearance between the ring and the barrel: too small and the barrel tends to jam, too large and vibration is set up between the ring and the barrel. The whole arrangement can be disadvantageous for barrel cooling, especially if the ratio of barrel length to cradle length is low.

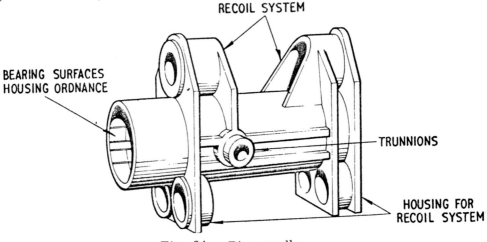

Fig. 34.   Ring cradle

THE SADDLE

The saddle is usually connected to the basic structure by a central pivot at its base. The pivot must be capable of withstanding the sheer stress applied when the gun is fired. The trunnion bearings in the saddle permit easy movement of the cradle in elevation and depression, but at the same time the capsquares provide for firm retention of the trunnions in their bearings, see Fig. 35. The saddle must not be allowed to tilt in the saddle support during the recoil cycle or

while travelling.   This is usually achieved by some form of holding-down device.
Nevertheless the saddle must be able to move easily in the horizontal plane when
the gun is traversed.   Some gun designs do not require any saddle whatsoever.
Such equipments are those with cross-axle traverse, in which the trail supports
the cradle and moves across the axle on traverse.   This design is not used for
modern close and depth support artillery systems.

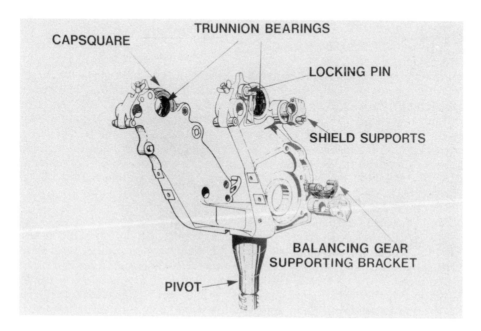

Fig. 35.   Saddle of a 105 mm Pack Howitzer L5

STABILITY

The Problem

Before discussing recoil mechanisms it is appropriate to consider the overall
question of gun stability.   Gun designers try to ensure that when a gun fires the
recoil energy produced is absorbed in a way that permits the carriage or mounting to remain stable.   A gun can be regarded as stable if the main supports of
the carriage or mounting remain stationary on recoil and run-out.   "Run-out" is
the term used to describe the action of the recoiling mass as it returns to its
original position after firing.   The term "counter recoil" is sometimes used instead of run-out.   As we shall see, if there are no limitations placed on the
weight of the gun the problem of stability is not difficult to solve.   In reality, however, the cry for high performance guns that are light enough to retain good
levels of strategic, tactical and battlefield mobility imposes conflicting requirements if the gun is to be regarded as stable.

If a gun is not stable the fire support provided will be adversely affected in several ways. The gun detachment will quickly realise that they must place them-selves in a safe position before the gun has fired. This will degrade the rate of fire and tend to make the detachment lose confidence in their equipment. Gener-ally the workload of the detachment is increased by the need to reposition the gun before firing the next round. Furthermore, there will be a risk of damage to the equipment too and even if this does not occur, accuracy and consistency will be affected.

The Components of Stability

The main components of stability are the weight of the gun and its trail length, the height of the trunnions and "trunnion pull". Trunnion pull is the force exerted at the trunnions by the recoiling parts. It acts parallel to the axis of the bore in the direction of the recoil path. All other things being equal, the heavier the gun, the longer its trails and the lower its trunnion height the more stable it will be. Nevertheless, the magnitude of the recoil energy to be absorbed is such that un-less a recoil mechanism is employed to reduce the recoil energy it becomes impossible to keep the gun stable merely by increasing weight or trail length, or by reducing trunnion height.

From Fig. 36 it can be seen that the gun will remain stable if $R < \dfrac{WL}{H}$ or $WL > RH$. A further consideration is the direction of R in the horizontal plane. Since the trails are the components of the carriage through which the recoil for-ces, which have not been absorbed by the recoil mechanism, wheels or supports, are transferred to the ground. The overturning moment produced by recoil be-comes critical if the line of recoil passes outside the "foundation figure" for the equipment, the foundation figure being the envelope determined by joining the outermost ground contact points with a straight line.

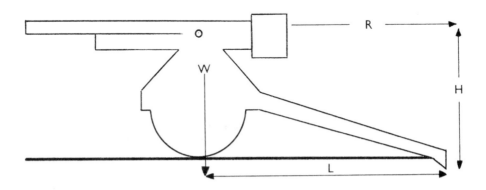

Fig. 36.   Main factors affecting the stability of a carriage

If the line of recoil is outside the foundation figure there will be a value of trun-nion pull above which the gun becomes unstable. Although a gun will tend towards overturning when it is unstable the available force is usually insufficient to do this. For practical purposes the gun can be regarded as unstable if one of the wheels or a trail leg lifts off the ground.   Gun designers seek to contain the direction of recoil within the foundation figure by limiting the traverse of the gun. The result is that to engage targets beyond the limits of traverse the trails have to be repositioned.

To determine the total resistance to be provided by the recoil mechanism in a carriage, the moment of the weight of the gun and carriage taken with respect to the point where the trail is supported by the ground must be greater than the moment of the resistance to recoil at the same point.   Assuming that the gun is on level ground and at zero elevation the limiting value of the total resistance to recoil can be expressed as follows:

Let Ws = weight of the static part of the gun acting at their centre of gravity 1.

Wr = weight of the recoiling parts acting at their centre of gravity 2.

R = the total resistance to recoil.

Lr = the length of the lever arm of Wr.

Ls = the length of the lever arm of Ws.

H = height of recoiling parts above spade centre of pressure.

d = the distance recoiled at any point during recoil.

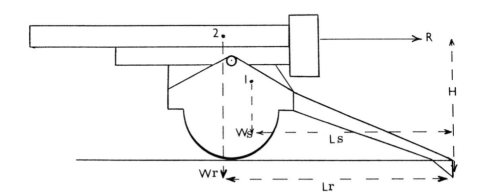

Fig. 37.   Resistance to recoil in a carriage

Assuming that the gun can be regarded as unstable when the wheels lift off the ground during recoil, the value of R if stability is to be maintained can be obtained by equating the moments in Fig. 37 as follows:

$$R = \frac{WrLr + WsLs}{H}$$

When the gun has recoiled through a distance (d) the lever arm of Wr becomes Lr-d and the limiting value of R is then:

$$R = \frac{Wr(Lr-d) + WsLs}{H}$$

From these two expressions it can be seen that the value of R within which stability is maintained at the commencement of recoil is reduced as the recoiling parts move further back and the moment about the recoiling parts decreases. Hence the value of R that ensured stability at the beginning of recoil may be insufficient to prevent the wheels lifting on further recoil.

This information is the basis of calculations used to indicate the permissable length of recoil for a given value of R, or if R can be made to vary, the extent to which recoil length can be reduced. If these calculations are not correct the recoil length could be either unnecessarily long and degrade the rate of fire or too short for the maintenance of stability. Even if the degree of instability is only slight, the effect on the time taken to re-lay and on the confidence of the detachment in the equipment can be detrimental.

If the recoil length is such that the gun remains stable at zero elevation then it will remain so at higher elevations. This fact is used to advantage in gun design and many modern gun carriages are provided with a method of reducing the length of recoil automatically as the gun is elevated. Other equipments take advantage of the increased stability at higher elevations in a different way by firing higher charges within a specified minimum limit of elevation.

The requirement for modern guns to have a good maximum range usually means that projectiles must be fired at high muzzle velocities. The additional stability problems created by increases in muzzle energy can often only be surmounted by adding to the total equipment weight. In some cases, the addition of extra weight may be either impossible or unacceptable because of mobility constraints. When faced with this impasse, the only alternative is to boost the projectile in flight or reduce the amount of drag on the projectile.

There are two other factors to be considered in gun stability: "jump" and the tendency of the barrel to rotate about its axis. Jump is defined as the displacement between the axis of the bore before firing and the axis of the bore after firing. On firing the force produced by the charge is exerted in the direction of the axis of the bore. Unless the centre of gravity of the recoiling mass remains on this line a couple is set up resulting in movement of the recoiling mass in the plane containing the axis of the bore and the centre of gravity. Jump can be predicted and corrections applied to gun data to ensure that the effects on accuracy and consistency are minimised. Additionally, gun designers endeavour to ensure that the recoiling parts and the centre of gravity remain on the same plane.

The tendency of the barrel to rotate about its axis is caused by the rotation of the projectile as it moves through the bore. The tendency is for the barrel to rotate in the opposite direction to the direction of rotation of the projectile. It can be expressed as a couple of magnitude:

$$\frac{DT}{2} \quad \text{where} \quad \begin{aligned} D &= \text{Diameter of the bore.} \\ T &= \text{Thrust on the rifling.} \end{aligned}$$

If the gun is to be stable this couple has to be balanced by one of equal moment.

Reducing Trunnion Pull

A reduction in trunnion pull can be turned to advantage by altering other gun design parameters while still maintaining a stable gun. For example, the total weight of the equipment can be produced and the trail length can be shortened. Trunnion pull can be reduced by the use of a muzzle brake, increasing the weight of the recoiling parts, or by increasing the length of recoil.

Muzzle brakes have already been discussed in Chapter 4 but to recapitulate, their effect is to decrease the rearward momentum of the recoiling parts by deflecting the gases moving up the bore behind the projectile. In the process the muzzle brake is pushed forward thus reducing trunnion pull.

The reason why trunnion pull can be reduced by increasing the weight of the recoiling parts can be explained as follows. According to the law of conservation of momentum:

$$M_p V_p = M_r V_r \qquad \text{where:} \quad \begin{aligned} M_p &= \text{Mass of projectile.} \\ V_p &= \text{Velocity of the projectile.} \\ M_r &= \text{Mass of recoiling parts.} \\ V_r &= \text{Velocity of recoiling parts.} \end{aligned}$$

$$\text{or} \quad V_r = \frac{M_p V_p}{M_r}$$

The kinetic energy (E) of the recoiling parts can be expressed:

$$E = \tfrac{1}{2} M_r (V_r)^2$$

Substituting for V_r

$$E = \tfrac{1}{2} \frac{M_p^2 V_p^2}{M_r}$$

It follows, therefore, that if the mass of the recoiling parts is increased less energy will be transferred to them with the result that trunnion pull will be re- duced. It should be noted that the use of a muzzle brake also increases the mass of the recoiling parts consequently producing a reduction in trunnion pull additional to that gained by its function in deflecting gases.

If the length of recoil is increased there is a decrease in the force acting on the trunnions: energy being the capacity to do work and work being the force applied multiplied by the distance over which it acts. If other means of improving stability cannot be exploited any further then lengthening recoil is an option. The disadvantages of this approach are that it reduces the rate of fire and could pre- sent other problems including restrictions in the safe working area behind the breech that can be occupied by the detachment at all times. For this reason increasing recoil length as a method of improving stability has greater applica- tion to towed equipments. For SP systems the overall weight of the equipment is the important factor in providing stability and designers will usually seek to keep recoil length as short as possible to relieve the constraints on the working area for the crew.

RECOIL SYSTEMS

All guns have recoil mechanisms to stop the rearward movement of the recoiling mass within a suitable distance. Recoil mechanisms also perform other allied tasks which are: to return of the ordnance to its firing position; to control of the ordnance run-out, or counter recoil; and to hold the ordnance in the run-out position throughout the complete range of elevations available to the gun.

The recoil cycle is as follows. When the charge is ignited a rapid rise in gas pressure begins; however, initially, the gun and projectile do not move. The projectile begins to move when the pressure is high enough to move it (Shot Start Pressure). At the same time recoil commences. Recoil continues until after the projectile leaves the muzzle and the gas action finishes. The run-out part of the recoil sequence then commences. The final stages of the run-out are more con- trolled and slower so that the gun is eased into the firing position, the carriage or mounting is not jarred and the original bearing and elevation of the barrel is not disturbed. The recoil cycle is completed when the gun has stopped moving. The main components of the recoil mechanism are: the "buffer" (sometimes called the "recoil brake"); "the recuperator" (sometimes called the "counter-recoil buffer"). There are two additional components that may or may not be used de- pending on the design of the major components already mentioned and the type of mounting employed. They are control to run-out arrangements and cut-off gears to vary the length of recoil.

The Buffer

The purpose of the buffer is to control the rearward movement of the recoiling parts. In the earliest systems buffers were of the friction type; however, modern buffers are hydraulic. Figure 38 shows an example of how a simple buffer might work.

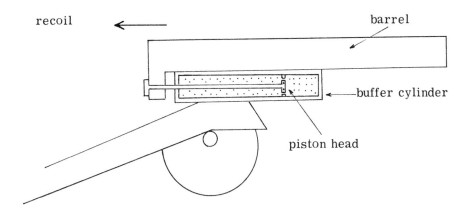

Fig. 38. Hydraulic buffer action

When the gun recoils the piston rod attached to the recoiling parts draws the piston head through the oil in the buffer cylinder. Oil is displaced through aper-tures from one side of the piston head to the other. The size of the apertures is such that the oil cannot pass through quickly enough to relieve the pressure behind the piston head. The ability of the buffer to check recoil lies in the pres-sure produced in this manner. The pressure retards the motion of the piston head, the recoil energy being converted into the kinetic energy of the moving oil and then finally dissipated as heat.

The simple diagram in Fig. 38 does not show how the velocity of recoil can be decreased uniformly. The work done by the oil in bringing the piston head to rest is equal to the resistance offered by the oil to movement through the aper-ture. If the size of the aperture can be made to vary so that it widens when the velocity of the piston increases and narrows as the velocity decreases the hyd-raulic pressure in the cylinder can be kept constant throughout recoil. Similarly, the apertures sizes could be varied to produce whatever levels of pressure are desired.

There are many different buffer designs to vary the oil flow. It can be made to flow through or around the piston head and can be varied by the shape of the cylinder walls, the shape of the piston head or the use of tapered rods or slots along the cylinder walls, just to name a few of the techniques used.

If necessary the length of recoil can be varied to give long recoil at low angles of elevation to achieve stability or short recoil at high elevations to prevent the breech from striking the ground or mounting. Some heavier mountings do not need the additional length of recoil for stability at lower angles of elevation and employ short recoil lengths at all elevations. The mechanical arrangements for variable recoil involve linking the oil flow control in the buffer to the elevation of the gun.

The Recuperator

The recuperator is the mechanism that returns the recoiled parts to their original position and holds them there until the next round is fired. An example of a simple form of recuperator is shown in Fig. 39.

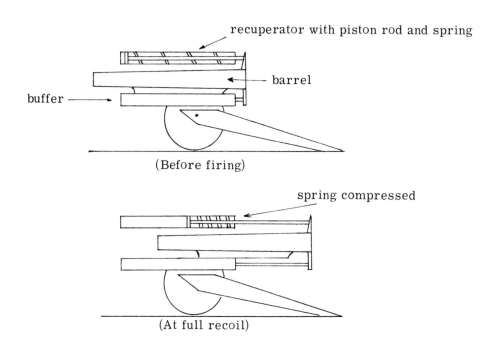

recuperator with piston rod and spring

barrel

buffer

(Before firing)

spring compressed

(At full recoil)

Fig. 39.   Recuperator

In the example shown the recuperator consists of a cylinder containing a spring and a piston. As the gun recoils the end of piston attached to the ordnance is drawn to the rear. By this action the piston head compresses the spring. At the end of recoil the spring expands and pushes the piston forward which in turn takes the barrel with it back to the run-out position. There are many different types of recuperator some employing springs, some employing compressed gas, some employing a combination of both. All of them work on the principle of using recoil energy to compress springs, gas, or both and then rely on the stored energy to restore the ordnance to its firing position. So far the buffer and the recuperator have been described as separate cylinders. It is possible to combine the functions of both into one cylinder.

Control of Final Run-Out

As mentioned earlier the final stages of run-out should be smooth and controlled. There are several means of achieving this effect. It can be incorporated into the design of the buffer or it can be a completely separate "control to run-out device" (sometimes called a counter recoil buffer). The control to run-out device, what-ever the technique employed, is simply an arrangement that controls run-out by slowing down the movements of the buffer piston as it returns to its original position. An example of one approach to control to run-out is in Fig. 40.

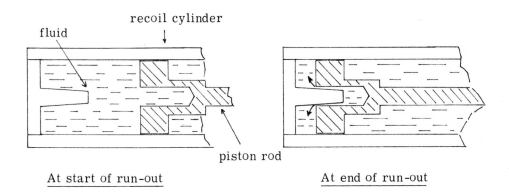

recoil cylinder

fluid

piston rod

At start of run-out   At end of run-out

Fig. 40.   Control to run-out

SOFT RECOIL

The artillery systems mentioned so far have highlighted a wide variety of different characteristics. Nevertheless they all have one thing in common: the barrel is stationary before the gun fires. Although that may appear to be a statement of the obvious there is a design concept that does have the barrel in motion before it is fired. Moreover, the potential advantages that could accrue from this basic, yet simple, change in the normal sequence of events on firing are worthy of men-tion. Many modern small arms use the concept of having their working parts held to the rear under tension. When the trigger is pulled the working parts are driven forward the weapon fires and the working parts are driven to the rear where they are held by a catch or a sear for the next shot. The rate of fire is much faster than with a conventional bolt action rifle and the recoil is less. A similar technique can be applied to larger calibre weapons and its adaptation to artillery weapons is known as soft recoil. The principle of soft recoil is illus-trated in Fig. 41.

As mentioned earlier the soft recoil principle is not new, at least as far as small arms are concerned. Perhaps surprisingly its application to artillery goes back many years to the French 2.65 inch Ducrest mountain gun issued in 1912: a gun that was not exactly a startling success. A modern example of a soft recoil gun, the 105 mm M204, was developed in the United States during the 1970s but has not been introduced into service, mainly for reasons of calibre.

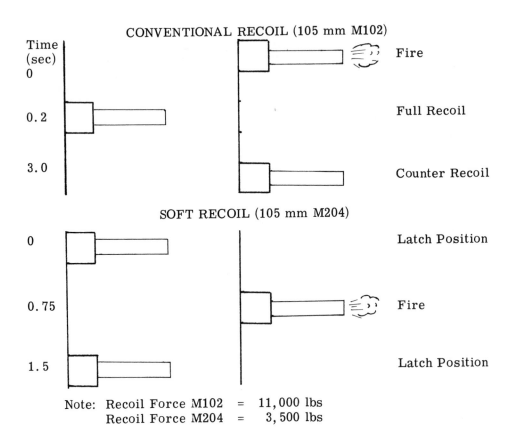

Fig. 41.   Soft recoil principle

The main advantages of a soft recoil artillery weapon are the reduction in trunnion pull and the decrease in the time taken for the recoil cycle.   The reduction in trunnion pull can be as high as 75% and improves gun stability to the extent that for the M204 trails and spades were not deemed necessary.   The gun is stable at any charge through 360° without any need to reposition the mounting.   A high rate of fire is possible at all angles of elevation and the detachment can work safely and unrestricted behind the breech.   This feature would be particularly useful for SP mountings where crew space is always at a premium.   Nevertheless the application of the soft recoil principle to artillery poses some unique problems for gun designers.

It is obvious that if the gun misfires there must be some means of arresting the forward motion of the barrel before it tips the gun over on its muzzle.   First of all some form of forward buffer system is needed to stop the barrel at the front of the cradle.   Secondly some means of returning the barrel to the latch position must be included in the design of the weapon.   Both of these requirements make the occurence of a misfire a little more complicated for a soft recoil gun than for a conventional gun.

Fig. 42.   M204 105 mm Howitzer

Range: 11500 metres

Weight: 4000 lbs

Rate of Fire: 10 rounds per minute

Note: This equipment was never introduced into service

Returning the barrel to the latch position would not only be a problem in the event
of a misfire but also poses difficulties with variable charge, or zoned, ammunition. At higher charges the recoiling parts need to travel further forward before
firing to balance the greater recoil. If the barrel is allowed to move too far forward the recoil will be insufficient to return the recoiling parts to the latch
position. The solution is to fit a velocity sensor to fire the gun when the correct
run-up speed for each charge has been attained or to fire at a predetermined
length of run-up. Clearly the likelihood of incorrect drills on a soft recoil equipment deserves special attention. In the worst case the velocity sensor could be
incorrectly set for the lowest charge but in fact the highest charge is fired,

resulting in the barrel overshooting the latch on recoil. To cater for this contin-
gency a buffer is required.

A further problem with soft recoil is the variation in ignition delay on firing.
Delay in ignition with a conventional gun is not important; however, with soft recoil it is critical. Delays as little as 30 milliseconds can necessitate the need for
a coast distance of 10 inches on the M204. The requirement for a coast distance
reduces the available recoil length back to the latch position. For calibres above
105 mm the ignition delay needs to be less than about 100 milliseconds for a soft
recoil system to be practicable.

BALANCING GEARS

As mentioned earlier in this chapter, a balancing gear is used when there is an
out of balance moment caused by the trunnions not being located at the centre of
gravity of the elevating mass. Although the addition of a balancing gear adds to
the complexity, weight, and sometimes silhouette of the gun, there are advantages
in designing the gun with "rear trunnions" or trunnions behind the centre of
gravity of the elevating mass. Rear trunnions permit a greater percentage of the
ordnance to be mounted forward of the area used by the detachment when loading
and firing the gun. This space is valuable on any equipment, but more so for an
SP mounting. Because the rear face of the breech mechanism is higher from the
ground at large angles of elevation, loading is easier and the need for variable
recoil is less. It will also be easier to design for traverse through 360° because
the breech will be less likely to foul the carriage or the mounting. Rear trunnions
may also aid concealment by permitting a lower silhouette.

The assistance provided by a balancing gear to overcome muzzle preponderance
varies with the cosine of the angle of elevation of the barrel (see Fig. 43).

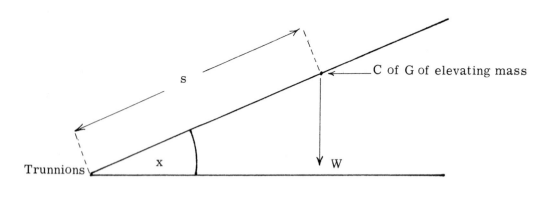

$$M = Ws \, Cosx$$ Where: M = out of balance moment
W = weight of elevating mass
s = distance from C of G to trunnions
x = angle of elevation

Fig. 43.   Out of balance moment

A balancing gear can operate by tension or compression with the force provided by metal springs, torsion bars, or a combination of spring and gas (pneumatic spring). Regardless of whether the balancing gear is a tension or compression type, the springs in the gear itself are usually loaded by compression. The difference between the two types of balancing gears can be explained as follows. In a tension type the gear arm attached to the basic structure, or saddle, is in tension and the gear spring is compressed (see Fig. 44). In a compression type gear both the arm attached to the basic structure and the balancing gear spring are under compression (see Fig. 44).

Metal springs are simple and reliable, with the compression spring type of balancing gear being easier to manufacture. Tension springs, however, produce better results for elevations over 45°. The advantage of a pneumatic spring is that, for a given task, it can be made smaller. Its disadvantage is that if the pneumatic seal is broken it will not function.

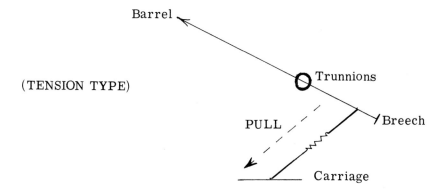

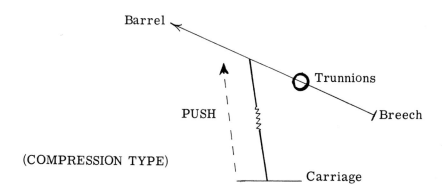

Fig. 44.   Types of balancing gear

ELEVATING AND TRAVERSING GEARS

Elevating Gears

The "elevating mass" comprises the ordnance the recoil mechanism and the
saddle. It is linked to the saddle by the elevating gear, the purpose of which is to
transmit and control the movement of the elevating mass relative to the saddle.
The main requirements of an elevating gear are that it should be conveniently
located; simple and easy to operate; non-reversible, so that the gun cannot elevate and depress of its own accord; and have sufficiently fine control to enable the
elevation of the barrel to be adjusted to the required angle. When the gun is fired
the turning moment on the recoiling mass imposes stresses on the elevating gear;
therefore, the ideal location for the elevating gear is in the same vertical plane
as the axis of the bore. Occasionally it is mounted to one side, but in such cases
it is duplicated at an equal distance on the opposite side. "Cradle clamps" are
provided on some equipments to prevent stress being placed on the gears during
travelling. A cradle clamp is a means of locking the cradle to the basic structure to ensure that it remains rigid as the carriage or mounting is moved across
uneven ground.

An elevating gear can be a simple gear train operated by a handwheel, a remote-
control power-driven mechanism, or an electric hydraulic mechanism controlled
by a handwheel. A simple gear train is the most common form of elevating gear
and 3 different types are used: worm and segment; nut and screw; or arc and
pinion. Some larger calibre guns cannot be easily loaded at large elevations and
"quick-loading gears" are provided to bring the elevating mass down to a convenient loading angle and then return it to the firing position. A quick loading
gear enables the gun to be placed in the loading position without affecting the
sights, hence loading and "laying drills" can proceed concurrently. Laying
drills are those actions concerned with applying data to the sights and pointing
the barrel at the required bearing and elevation. As an alternative some large
equipments have 2 elevating gears: a coarse gear for rapid elevation and a fine
gear for accurate final laying. Electric and electric-hydraulic elevating gears,
however, provide the best means of solving the problem of loading and laying
large calibre equipments for high rates of fire. Because of the power requirements this type of elevating gear is mainly reserved for carriages or mountings
with an integral power source at the equipment, such as SP mountings.

Traversing Gears

A traversing gear is a mechanism for moving the superstructure in the horizontal plane. It does so by transmitting motion to the saddle to position it in the
horizontal plane relative to the basic structure. The requirements of a traversing gear are similar to those already mentioned for elevating gears. As with
elevating gears, the types of traversing gear include worm and segment, nut and
screw, as well as arc and pinion. A further type, "rapson nut and screw", is
also used. Rapson nut and screw is a form of nut and screw gear using a recir-
culatory ball race. Traversing gears can be manually operated or power driven,
with power driven versions usually having the facility for manual operation as

well. Hydraulic or electric systems can be used for power traverse and in some
systems power assistance is only used in situations where the effort required by
manual means is excessive for example when the equipment is on a slope.

In most modern guns the saddle and the basic structure are connected by a central
pivot at the base of the saddle. It is essential for precision in traverse that any
vertical movement of the saddle is prevented and that the horizontal movement is
restricted to that intended. The bearings used for the pivot must also be able to
cope with the firing loads and the traverse rate for the gun. Dissimilar metals
or steels of differing hardness are used to help cope with this problem and for
high loads and traverse rates, ball or roller bearings are employed.

Some artillery weapon systems have the capability for traverse through 6400 mils
without having to reposition the basic structure of the carriage or mounting. The
requirement for 6400 mils traverse has to be carefully weighed against the
stability problems this may impose. Some modern SP mountings have centrally
mounted turrets to facilitate 6400 mils traverse; however, the penalties imposed
are additional weight and less than ideal access to the crew compartment for
ammunition replenishment. Towed equipments with 6400 mils traverse sometimes
have quick release mechanisms so that the superstructure can be disengaged from
the traversing mechanism and rotated quickly to the required bearing. The quick
release mechanisms only allow for the barrel to be pointed in the approximate
direction required and the traversing mechanism is re-engaged for final adjustment. The Soviet 122 mm D30 Howitzer has this facility.

It is common practice for more conventional towed equipments to have limits
placed on traverse and these usually take the form of mechanical stops on the
traversing gear. The limits imposed are not always for reasons of stability alone,
but also to prevent fouling of the recoiling parts and the traversing mass with the
basic structure, especially at high angles of elevation. Variable limits of traverse are possible in equipments where the height of the trunnions and the axle
width can be altered. An example of this is the Italian 105 mm Pack Howitzer
which has a traverse of 644 mils in its normal firing configuration and a traverse
of 996 mils in low, anti-tank position.

The limits of traverse imposed by the design of a gun is very much a function of
its role and the facility with which the basic structure can be repositioned for the
engagement of targets beyond the limits of traverse. The shorter the range of
the equipment and the shorter the range at which targets appear the more critical the limits of traverse become. For example, two targets 1500 metres apart
at a range of 3000 metres from the gun require a change in traverse of 500 mils
when they are engaged in rotation. If the same targets were engaged by a gun
from a range of 30,000 metres the displacement in terms of traverse would be
50 mils. Similarly, a gun employed in the anti-tank role needs wide traverse
limits to cope with the lateral movement of targets at comparatively short ranges.

Fig. 45.   105 mm Pack Howitzer in normal firing position

Fig. 46.   105 mm Pack Howitzer in anti-tank position

SIGHTS

Indirect Fire

The term "indirect fire" is used to describe fire applied when the target is not visible from the gun; or if it is, the direct vision link between sight and target is not used. The process of adjusting the gun for line (direction) and elevation is called "laying". The laying of a gun involves two separate operations, though in practice they are often closely linked: laying for line and laying for elevation. The purpose of the sights is to provide a means of positioning the barrel at the correct line and elevation to hit the target. Although the elevating and traversing mechanisms are the means of moving the barrel in line and elevation, the sights are used to measure the angles through which the barrel must be moved before it is correctly positioned for firing.

A simple explanation of how a gun is laid for line is as follows. When a gun is initially deployed it is emplaced with its barrel pointing in a pre-determined direction, usually the bearing to the centre of the area in which targets are likely to be engaged. The sight used to lay the gun for line is supported in a bracket mounted either on the cradle or pivoted to the saddle and linked to the cradle. At the time the gun is deployed the relationship between the barrel, the sight and a "gun aiming point" is fixed and recorded. The gun aiming point is simply a reference point, natural or artificial, visible through the sight. Figure 47 depicts the situation just described with the gun barrel pointing due north and the sight directed at a gun aiming point due west. Theoretically the gun aiming point could be in any direction although the selection of its location may often be limited by obstructions caused by parts of the gun itself or by any other obstruction to the field of view around the gun.

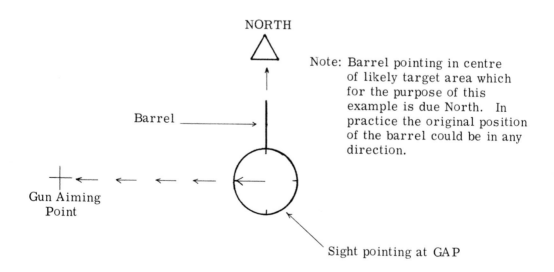

NORTH

Note: Barrel pointing in centre of likely target area which for the purpose of this example is due North. In practice the original position of the barrel could be in any direction.

Barrel

Gun Aiming Point

Sight pointing at GAP

Fig. 47. Relationship between gun aiming point and sight
(before sight setting)

Let us now assume that a target due east of the gun is to be engaged. Data is applied to the sight so that the horizontal angle through which the barrel must be moved can be measured. The effect of applying this data to the sight is to rotate it independently of the barrel, through a horizontal angle equal but in the opposite direction to that through which the barrel must eventually be traversed. At this stage the barrel has not been moved but the sight has been offset from the gun aim-ing point.

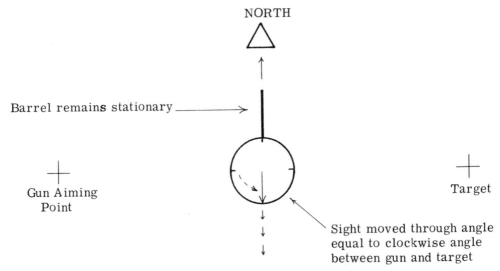

Fig. 48.   Relationship between gun aiming point and sight
(after sight setting)

The next step is to traverse the gun until the sight is once more pointing at the gun aiming point. The barrel is now laid at the correct bearing to hit the target (see Fig. 49).

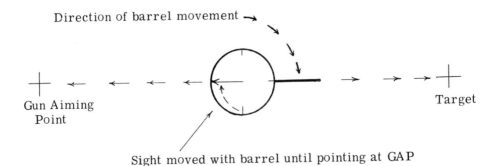

Fig. 49.   Relationship between gun aiming point and sight (lay complete)

In practice the gun is not aimed in the precise direction of the target but is offset
to allow for the effects of meteorological conditions and "drift". The corrections
for the prevailing meteorological conditions are incorporated in the data applied
to the sight. Drift is the lateral deviation of the projectile during its flight resulting from the spin imparted to it by the rifling in the bore of the gun. For a shell
spinning clockwise the drift will be towards the right. This phenomenon is caused
by the fact that a spinning projectile tends to maintain the same longitudinal axis
as it had at the muzzle. As the trajectory curves there is a degree of yaw or
horizontal angular displacement between the tangent to the trajectory and the axis
of the projectile. The centre of gravity of the projectile remains on the trajectory; however the effect of rotation makes the tip of the projectile move off to the
right. The result is a displacement of air to the left and a consequent drift of the
shell to the right. For a given projectile and charge, drift increases with the time
of flight of the projectile. Some sights compensate for drift. Alternatively, the
necessary correction to line for drift can be included in the data to be applied to
the sights. Yaw is not the only effect on the projectile that will cause it to drift.
Magnus effect and the rotation of the earth are the other components of drift;
however, yaw is the most significant component.

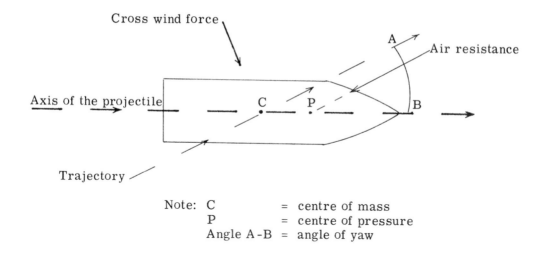

Note: C   = centre of mass
       P   = centre of pressure
       Angle A-B   = angle of yaw

Fig. 50.   Drift of projectile in flight

When firing data is computed it is assumed that the gun is on a level platform.
Often the ground may not be even and consequently the trunnions may not be level.
If the trunnions are not level there will be a line error in the direction of the
lower trunnion because the gun elevates at right angles to the axis of the trunnions. Errors caused by the lack of level of trunnions are avoided by the use of
levelling mechanisms to ensure that the sight remains vertical and the gun can be
laid accurately. Sights that allow corrections for drift and lack of level of trunnions to be applied automatically are called compensating and reciprocating
sights. The recent trend is to use sights that allow for lack of level of trunnions
but not for drift.

Automatic correction for drift results in complexity in the design of sights for
multi-charge equipments firing a variety of projectiles.

Guns also need sights for the application of angles of elevation so that the projectile will achieve the desired range. The angle of elevation applied to the gun has
two main components, "tangent elevation" and "angle of sight". Tangent elevation
is the vertical angle at which the barrel must be laid to achieve a given range for
a given charge, assuming that the gun and target are on the same horizontal plane.
Angle of sight is the angle between the horizontal plane and the line of sight to the
target (see Fig. 51). If the target is higher than the gun the elevation must be
increased and vice versa if it is lower. The algebraic sum of these two components of elevation is called quadrant elevation.

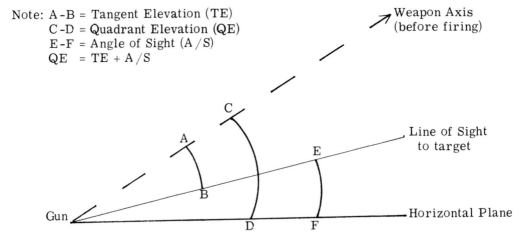

Note: A-B = Tangent Elevation (TE)
      C-D = Quadrant Elevation (QE)
      E-F = Angle of Sight (A/S)
      QE = TE + A/S

Fig. 51.   Ballistic angles

For a given range the tangent elevation is not the same for all target heights. In
other words the trajectory cannot be swung rigidly above and below the horizontal
and a correction to tangent elevation is normally needed when an angle of sight is
applied. This correction for what is termed "non-rigidity of trajectory" is
allowed for in the data applied to the sight.

The design of some sights compensates for variations in muzzle velocity due to
wear. Sights with this facility are called calibrating sights. The most common
form of sight for elevation is the "rocking bar sight". The rocking bar sight tilts
through an angle in the vertical plane when elevation data is applied to it. The
barrel and the sight are then elevated (or depressed) together and when the sight
is level once more the gun is at the elevation required. The reference for elevation is usually a longitudinally mounted spirit level bubble or a set of pointers.
In some sighting systems tangent elevations and angle of sight are set on the gun
independently by different members of the gun detachment. This system is known
as independent line of sight (or sighting) and its purpose is to divide the respon-
sibilities in laying for elevation thus permitting increased rates of fire.

The trend in modern sighting systems is to allow for lack of level of the trunnions
only, with the corrections for drift and muzzle velocity being computed separately

and incorporated in the data applied to the sight. Alternatively the correction for muzzle velocity can be calculated by the use of a "gun rule". A gun rule is a type of slide rule kept at the gun. It is used to deduce the range or elevation for a particular gun corrected for its particular muzzle velocity at whatever charge is being used. The reason for the trend away from compensating and calibrating sights are the expense, complexity and size of such sights together with the use of computers in the command post that simplify the calculation of corrected data for application to the sights.

Direct Fire

The purpose of sighting systems for direct fire is generally the same as for indirect fire: to ensure that the trajectory intersects the line of sight at the target. Laying the gun for direct fire can be achieved by using the normal indirect fire dial sight or by using a separate direct fire telescope. Direct fire is only used at comparatively short ranges and when the target is visible from the gun. In the past some gun systems have been specifically designed for direct fire in the anti-tank role. Nowadays gun systems for close and depth support are often equipped with sighting systems that permit direct fire; but very much for secondary tasks such as the defence of the gun position against armour or infantry. Sighting systems for anti-tank guns do not normally compensate for drift or lack of level of the trunnions because of the short ranges involved. Corrections for muzzle velocity and jump are allowed for in the zeroing of the sights.

The main problems in direct fire sighting are to ensure that some arrangement is made for the engagement of moving targets and that the sight or telescope used is zeroed for the type of ammunition to be fired. Some gun systems have special ammunition for direct fire anti-tank engagements, others do not and rely on the use of a normal HE projectile. The lateral movement of the target is overcome by displacing the line of sight from the axis of the bore so that when the sight is laid on a moving target the barrel is pointed ahead of the target by an amount that will ensure a hit. The amount of aim-off or "lead" that is applied depends on the muzzle velocity of the gun as well as the speed and direction of the target. The sight or telescope used may be inscribed with a graticule pattern with horizontal lines for range and vertical lines for left and right leads. In some systems the tangent elevation can be applied to the sights as for indirect fire leaving the lay for line to be effected by the use of the graticule pattern. This approach is more common in systems that use a two-man lay: one for line and one for elevation. Other systems rely on a central laying method, with the laying mark always being the centre of the field of view of the sight or telescope and the necessary lead being applied by some mechanical arrangement to deflect the sight line to the right or left of the bore. In both systems the act of laying the sight onto the target automatically corrects for any difference in height between gun and target.

SECTION 2.   THE BASIC STRUCTURE

CARRIAGES

General

The basic structure of a carriage is made up of the saddle support, the trails and
method of articulation; platforms and spades; the wheels, axles, suspension and
brakes. The saddle support provides the foundation on which the saddle rests with
the horizontal movement between the two controlled by the traversing gear.

Trails

The trails are that part of the basic structure of a carriage through which the
forces of recoil are transmitted to the ground. The trails help to hold the gun in
the firing position and serve to connect the weapon with its prime mover or towing vehicle. The forward end of each trail is usually fastened to the axle or to
the saddle support. The rear end is usually fitted with some form of spade which
may be adjustable or even removable. When the gun is in the towing position the
trails are coupled to the prime mover by some form of towing attachment or by a
"lunette". The towing attachment is usually removable. A lunette is a mechanical arrangement attached to the rear end of the trail and designed to mate with
the towing hook on the towing vehicle. Lunettes are usually designed so that they
do not interfere with the action of the trail or spade when the gun is emplaced.
There are three types of trail: pole trail, box trail and split trail. These different trail configurations offer varying scope for traverse and elevation as well
as different penalties in weight.

The simplest and lightest type of trail is
the pole (or single trail) configuration.
The earliest examples of guns had single
trails. The top traverse available with
a pole trail is usually limited to about
150 mils. Any increase in the traverse
for this configuration would pose stability
problems. Furthermore, the elevation
is typically limited to less than 300 mils.
In the past pole trails have been used in
equipments where simplicity and light-
weight have been paramount, such as in
pack guns. They have also been used for
equipments with axle traverse. Nowadays this concept is regarded as unsuitable because of the limitations in elevation and traverse. For this reason there
are no modern guns with pole trails.

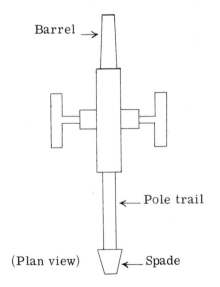

(Plan view)

Fig. 52.   Pole trail

A box trail is made up of two stabilising members joined at the rear or two side members and a rear cross member. The two side members can be bowed, splayed or parallel. A bowed or a splayed box trail permits a top traverse of about 200 mils with parallel sided box trails usually somewhat less. Box trails often have firing platforms to give suitable support in soft ground and facilitate further traverse through 6400 mils. The box trail configuration allows space for recoil and the movement of the breech in the vertical plane. Box trails have been favoured in modern British carriages whereas other nations, including the United States, seem to favour "split trail" towed guns and howitzer.

Fig. 53.   UK Light Gun (example of box trail)

As the name suggests a split trail has two stabilising members which are spread for firing as far as permitted by either the wheels or the trail stops. The design of a split trail gun usually allows for the trails to spread at an angle equal to or greater than the top traverse of the gun. The split trail configuration usually gives a greater top traverse than a box trail; however, it is sometimes more difficult to move the gun through 6400 mils. To reposition the trail of a box trail gun takes at least one man, whereas for a split trail it takes a minimum of two. Because the trails of a split trail gun are spread when the gun is in the firing position, the effective trail length is shorter than the actual trail leg length. Consequently the trail legs are longer and heavier than box trails. In addition, when the gun is fired at the extremes of its top traverse the firing stresses tend to be transmitted to one trail leg: a further reason for greater length and weight. Split trails permit freedom of movement of the breech in the vertical plane and good access for the detachment serving the gun. Cut-off gears may still be necessary to limit recoil when the gun is firing at high angles of elevation so that the breech

does not strike the ground. With some equipments recoil pits may have to be dug
beneath the path of recoil but this minor inconvenience is sometimes accepted
rather than designing the gun with a greater trunnion height. The individual trail
legs can be either in one piece or in sections. They may also be straight or
cranked. The Italian 105 mm Pack Howitzer is an example of an equipment that
incorporates variable split trail configurations (see Fig. 46 on page 98).

Articulation

Guns with box or pole trails can remain stable on rough ground because there are
only three points of contact with the ground. Split trail equipments have four
points of contact and on uneven ground four points of contact can only be maintained
by some form of articulation in the basic structure to permit sufficient freedom of
movement in the vertical plane. The means of achieving this is called the method
of articulation. In some armies the mechanical arrangement for achieving articulation is called the equaliser. There are many methods used for achieving
articulation. Most involve the use of pins, rocking arms or ball socket joints to
obtain the relative movement between the trail legs and the axle tree or the saddle
support, depending on the means used to connect the trails to the remainder of the
basic structure. Whatever method is employed, there must also be provision for
locking the articulation when the gun is being towed and some arrangement for
preventing the trails from closing when the gun is in the firing position.

Spades

Spades are normally fitted at the end of a trail leg to restrict movement of the gun
during recoil. Spades can be fixed, detachable or hinged. Fixed spades are permanently and rigidly fastened to the end of the trail legs. They can be a forging,
a casting, or of built-up construction. A detachable spade is one that is removed
from the trail for travelling and replaced for firing. Hinged spades are similar
to fixed spades except that they can be swung out of the way for travelling. Some
guns have two sets of spades: a hinged set and a further set for use in softer
ground. The problems in spade design stem from the sometimes conflicting requirements of transmitting horizontal firing stresses to the ground while remaining easily removable from the ground to facilitate movement of the trails.

A small spade transfers more energy per unit of ground contact area than a large
one; therefore, the soil is more likely to move than if a larger spade were used.
Conversely, a large spade is necessarily heavier, is more difficult to extract and
may be of little use in hard rocky ground. It is difficult, if not impossible to
design spades that are ideally suited to all conditions in terms of both spade area
and shape. The bearing strengths of different soils, or the loading they will withstand without general shear failure, varies greatly. For example, the loading
that firm gravel will withstand without failure is about five times greater than
soft clay. Additionally, the dynamic loadings imposed by a gun firing tends to
compact some frictional soils such as sands and gravels with the effect that the
amount of movement decreases with each round fired. On the other hand cohesive
soils such as mud and some clays do not compact easily to offer resistance to
movement. Furthermore some harder soils such as chalk may tend to crumble

as each round is fired. Little wonder, therefore, that the size and shape of spades varies so much.

Platforms

Platforms are usually fitted to equipments on which the trails cannot absorb residual firing stresses unaided. They are, therefore, commonly used with box trail configurations (see Fig. 53 on page 105). As was mentioned earlier in this section, platforms assist in transmitting to the ground the vertical components of residual recoil energy, help prevent the gun sliding backwards and provide a firm base for the wheels to run on when the trails are repositioned. The design of the platform must permit good ground adhesion throughout the range of soil conditions to be encountered. They are usually circular, with a rim on which the wheels can run. The rim is sometimes supported by spokes or the whole platform can be solid. In the firing position the platform is attached to the gun by wire rope stays or rigid brackets. The platform must be raised for travelling and before this can be done the gun is normally manhandled off the platform.

Wheels and Axles

The wheels on modern carriages are low pressure, pneumatic-tyred wheels usually of the "run-flat" type, which means that if the tyre is punctured it can still carry the weight of the equipment for some distance until repairs can be effected. The axle connecting the wheels of a gun carriage can be "underslung" or "overslung" meaning that the horizontal line between the wheel centre can be above the axis of the axle (underslung) or below it (overslung). An underslung axle permits a lower weapon silhouette and improves stability. Its disadvantage is that the elevation of the equipment may be limited. The greater the length of the axle and hence the greater the distance between the wheels, the greater the scope for top traverse and the more stable the gun when it is being towed. A wide track does have some disadvantages, the main ones being the greater effort needed to move the trails and the fact that the axle is more likely to foul obstructions on uneven ground. The wheel/axle configuration is usually a compromise based on the most likely tasks for the weapon. The Italian 105 mm Pack Howitzer has a variable wheel track and the axle can be either overslung or underslung. Such versatility is only practicable in light equipments. Moreover, it is usually achieved at the expense of increased cost, complexity, and reduced reliability. Some World War I vintage equipments such as the US M1918 155 mm Gun used removable track treads on their wheels to reduce the ground pressure of the equipment in soft going. Modern equipments do not normally employ this technique, relying instead on low pressure tyres with an aggressive tread to improve traction in soft going.

Brakes and Suspension

The brakes on most carriages are of the conventional brake drum and shoe or band type. Several different design concepts may be used to activate the brakes.

They can be operated from the towing vehicle by the use of an electrical or mechanical connection to the gun and operated by air, electricity or hydraulically. Some heavier equipments have separate safety breakaway systems that operate if the gun accidently becomes unhooked from the towing vehicle.

Besides the main braking system used for towing, carriages usually have hand operated brakes. Some equipments have one hand lever for both wheels, others have a lever for each wheel. The purpose of hand brakes is to help prevent movement of the gun during firing and to enable it to be parked on slopes. The hand brakes are also used to lock one or both wheels in position during manhandling over sloping or broken ground. The design of most carriages does not incorporate a suspension system, the pneumatic tyres providing the only improvement to the smooth travel of the gun behind the towing vehicle. The UK Light Gun is an exception to this general rule, having trailing arm suspension, with torsion springs and shock absorbers. The result is that this gun has an exceptionally good towing speed on roads and cross-country.

BASIC STRUCTURE OF MOUNTINGS

Mobile Mountings

As mentioned earlier in this chapter, a mobile mounting is an equipment that travels on wheels but, unlike a carriage, does not fire off them. When a mobile mounting is deployed the wheels are either removed or raised off the ground so that they do not support the equipment in any way. The main advantages of mobile mountings are the improved stability achieved by the fact that the equipment is supported by a solid mounting and the advantage that once the wheels have been removed or raised the gun's centre of gravity can be lowered. These stability advantages can result in greater accuracy and consistency. A further advantage is that with some mobile mountings the whole superstructure can be moved through 6400 mils without shifting the basic structure as is necessary with a carriage. Compared with carriages, mobile mountings are usually heavier and slower to emplace. The latter disadvantage is normally not critical as the extra time taken to emplace the equipment is not always a significant part of the total time taken for the drills to prepare a battery of guns for firing. There are three main types of mobile mounting suitable for close and depth support tasks: pedestal and stabilising girders; platform and stabilising girders; and platform and sole plate.

An example of the pedestal and stabilising girders type of construction can be seen in Fig. 54. The stabilising girders are designed so that stability can be maintained through 6400 mils traverse. There is no articulation between the pedestal and the girders, thus increasing the importance of a firm level site for the gun platform.

Fig. 54.   Mobile mounting
(pedestal and stabilising girders)

Fig. 55.   Mobile mounting
(platform and stabilising girders)

Figure 55 shows an example of the platform and stabilising girders type of mobile
mounting. In this type of construction the trails perform the same function as they
do with a carriage except that they are in contact with the ground throughout their
length. The wheels have been raised clear of the ground and the full weight of the
equipment is on the platform and stabilising girders. Articulation is usually not
provided with this type of construction. Spades are used to prevent movement
when the gun is fired and in some examples of this configuration additional spades
are attached to the basic structure between and outside the stabilising girders.

The platform and sole plate type of mobile mounting is used with heavier equipments. An example of this type of mounting is shown in Fig. 56. This method of
construction is sometimes called a turntable mounting. The design incorporates
a large traversing ring mounted on a supporting base. There is an out of balance
moment to the rear; however, the equipment can be easily traversed through 6400
mils by raising the rear with a roller and jack system. A sole plate at the rear
of the basic structure permits a limited amount of top traverse.

Fig. 56.   Mobile mounting
(platform and sole plate)

SP Mountings

The vehicle chassis provides the basic structure for an SP mounting. The saddle
is usually mounted in some form of traversing race and the equipment can have
limited top traverse, or 6400 mils traverse, depending on the vehicles design.

When the gun fires, movement in the horizontal plane is prevented by the comparatively heavy weight of the equipment and the length of track in contact with the
ground. Some SP equipments have rear spades to improve stability (see Fig. 57).
The requirement for an SP to fire in high angles of elevation often results in the
vehicle's silhouette being higher than would otherwise be necessary in order to
provide sufficient space for recoil. 6400 mils traverse can be achieved by providing the equipment with a turret and although this may often be desirable, the
positioning of the turret to balance the vehicle can make access for ammunition
replenishment much more difficult. Alternatively a limited arc of traverse can
be accepted and the engagement of targets beyond the limits of traverse achieved
by a design that permits the vehicle to pivot about one track. To protect the
vehicle's suspension system from being subjected to firing stresses, some SP
mountings are fitted with suspension lock-out arrangements.

Fig. 57.   SP Gun (with rear spade)

This chapter has reviewed the main design features of carriages and mountings.
There are many other features that are beyond the scope of this book and no doubt
engineering ingenuity will provide more in the future.

SELF TEST QUESTIONS

QUESTION 1   What is the difference between a carriage and a mounting?

Answer   ....

....

QUESTION 2   List the main components of the superstructure.

Answer   ....

....

....

QUESTION 3   Why are balancing gears fitted to guns?

Answer   ....

QUESTION 4   What is a cradle and how is it supported?

Answer   ....

....

QUESTION 5   Explain the term "run-out".

Answer   ....

....

QUESTION 6   List the main gun design parameters that determine its stability.

Answer   ....

....

....

....

QUESTION 7   Explain the term "foundation figure".

Answer   ....

....

QUESTION 8   Calculate the maximum values of resistance to recoil to ensure
stability of a gun, at the instant of firing at zero elevation, given
the following information:

a.   The gun is a towed carriage with a box trail.

b.   The weight of the recoiling parts is 2000 lbs and the centre of gravity of the recoiling parts is 160 inches from the spade and 45 inches above the spade centre of pressure.

c.   The weight of the basic structure is 5000 lbs and its centre of gravity is 115 inches forward of the spade.

Answer   . . . .

. . . .

. . . .

. . . .

. . . .

. . . .

. . . .

QUESTION 9   Why do gun designers design equipments for recoil lengths that vary with elevation?

Answer   . . . .

. . . .

. . . .

. . . .

QUESTION 10   Explain why trunnion pull can be reduced by increasing the weight of the recoiling parts.

Answer   . . . .

. . . .

. . . .

. . . .

QUESTION 11   What happens to the force acting on the trunnions if the length of recoil is increased and why?

Answer   ....

....

QUESTION 12 Name the 2 main components of a recoil mechanism.

Answer   ....

QUESTION 13 What is the purpose of a "control to run-out device"?

Answer   ....

QUESTION 14 Explain briefly what is meant by the term "soft recoil".

Answer   ....

....

....

QUESTION 15 What are the advantages and disadvantages of soft recoil?

Answer   ....

....

....

....

QUESTION 16 List the advantages of having rear trunnions on a gun.

Answer   ....

....

....

....

QUESTION 17 What is the purpose of a cradle clamp?

Answer   ....

QUESTION 18 Why are limits imposed on the traverse of some equipments?

Answer   ....

....

....

QUESTION 19   Explain the terms "direct fire", "indirect fire" and "laying".

                Answer   ....

                            ....

                            ....

                            ....

                            ....

                            ....

QUESTION 20   Define the term drift.

                Answer   ....

                            ....

QUESTION 21   What is a reciprocating sight?

                Answer   ....

                            ....

QUESTION 22   Using the diagram below indicate the following ballistic angles:
                angle of sight, tangent elevation, quadrant elevation and jump.

                Answer

Line of Departure

Axis before firing

Line of Sight

Horizontal

QUESTION 23   Explain the term "lead".

                Answer   ....

                            ....

QUESTION 24   List the main components or groups of assemblies that make
                up the basic structure of a carriage.

Answer   ....

           ....

           ....

           ....

           ....

           ....

QUESTION 25   Why are pole trails not used in modern carriages?

               Answer   ....

                          ....

QUESTION 26   List the advantages and disadvantages of split trails compared
               with box trails.

               Answer   ....

                          ....

                          ....

                          ....

                          ....

                          ....

QUESTION 27   What is an equalizer?

               Answer   ....

                          ....

QUESTION 28   Why are spades of different shapes and sizes used, sometimes
               even for the same equipment?

               Answer   ....

                          ....

                          ....

QUESTION 29   Do gun carriages have suspension systems or not?

               Answer   ....

QUESTION 30   What is a mobile mounting?

                Answer   ....

                           ....

QUESTION 31   What method is used to help protect the suspension system
                of an SP gun from damage when the gun recoils?

                Answer   ....

ANSWERS ON PAGE 192

6.

Mortars

GENERAL

The main characteristics of mortar systems compared with guns and rockets have already been covered in Chapter 3. Although some mortars are rifled and/or breech loaded, a conventional mortar is a smooth bore, muzzle loaded weapon without a mechanical recoil mechanism and fires a fin-stabilised bomb at elevations above 800 mils. The advantages of a mortar system accrue primarily from the comparative simplicity of the system. The main components of a mortar are the baseplate, the barrel, the bipod and the sights. Each of these components or groups of assemblies will now be discussed, together with related aspects of contemporary mortar design and employment.

BASEPLATES

Function

Mortar baseplates provide the means of support for the breech end of the barrel and transmit the force of recoil to the ground. Besides being robust enough to withstand the force of firing without being too large or too heavy, baseplates must also be easy to emplace and remove. In addition, the design of the baseplate should permit the degree of traverse needed in the concept of operations in which it is to be employed.

Ground

The general considerations for the design of spades on gun trails, discussed in Chapter 5, apply equally to mortar baseplates. In essence the design of the perfect baseplate for all ground conditions is difficult, if not impossible. Furthermore, there appears to be a paucity of information on the behaviour of different soil types under the dynamic loadings produced on the ground when a mortar fires.

Ideally a baseplate should settle quickly into a firm, stable position and thereafter not move at all. In practice this does not happen and movement varies from
an insignificant amount to comparatively large movement that will affect the consistency of the spread of mortar bombs at the target.

At present there is no practical, foolproof method of preparing the ground for a
mortar baseplate for use under operational conditions. Perhaps the exception to
this is when a mortar is emplaced in a comparatively static location, in which
case extensive preparation in the form of a concrete slab may be possible.
Nevertheless, such measures can be regarded as being contrary to the normal
practice for the employment of mortars as it presupposes that they will not move
often. This approach negates one of the prime advantages of mortar systems:
mobility. Further, a mortar firing regularly from a static location can be easily
detected by radar because of the weapon's high trajectory and the signature of the
finned mortar bomb. Consequently if the system's mobility is not exploited its
survivability may be degraded.

Past experience has shown that the preparation of a mortar baseplate position
without "bedding-in" is usually unsatisfactory. Bedding-in is the process of firing
two or more rounds to ensure that the baseplate is firmly settled into the ground
so that subsequent firing can be carried out without the baseplate rocking.
Usually the bedding-in procedure is carried out with a high charge fired at a high
elevation. Rounds that are fired before the mortar has been properly bedded-in
will tend to range short because of the amount of energy lost in recoil. Even
when the bedding-in drill has been completed there is a risk that further firing
may bring the baseplate into contact with a rock or some other obstruction in the
soil that may bear against one section of the baseplate, thus introducing a high
local loading on the baseplate. The result could be that the baseplate may pivot
or twist about the obstruction, causing instability. Even if this does not occur
the section in contact with the obstruction must be sufficiently strong to withstand
any additional loading.

When a mortar fires there is a horizontal and a vertical component to the ground
reaction on the baseplate. The magnitude of these components varies with the
elevation of the mortar barrel and can be explained as follows. Using the example
in Fig. 58 on the next page, if we assume that R is 40 tonnes force and that the
elevation is 1000 mils, then the horizontal component will be 22 tonnes force
and the vertical component will be 33 tonnes force. Obviously the horizontal
component will be reduced as the elevation increases; however, even at high
angles of elevation the baseplate must grip the ground sufficiently to withstand
the horizontal component. Baseplates are usually fitted with some form of shoe
or spike for this purpose, but regardless of the design some degrees of manual
preparation may also be necessary. The preparation usually takes the form of
digging-in or building up a suitable support by the use of sandbags.

If a mortar is fired below 800 mils the horizontal component of the ground reaction increases as the elevation decreases. The tendency for the mortar to slip
backwards on firing increases and although methods could be improvised to
reduce this tendency, the results achieved are usually unsatisfactory.

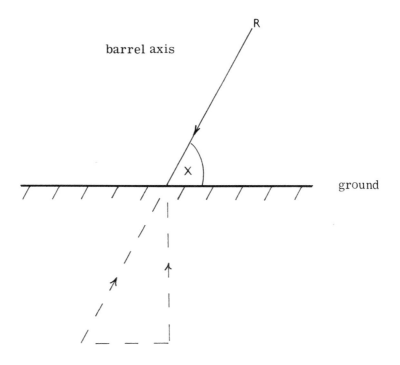

Fig. 58.   Components of mortar recoil

Where:

R   = Recoil Force

X   = Angle of Elevation

Horizontal Component = R Cos X

Vertical Component   = R Sin X

Some baseplates are designed so that when they are placed on level ground they slope at an angle. The angle is such that the recoil force is transmitted at approximately right angles to the baseplate and parallel to the axis of its spikes. The baseplate spikes in this concept usually vary in length with the shorter spikes being at the front. The disadvantage of this approach is that it precludes all round traverse. Nevertheless, it is a disadvantage that is shared, to a lesser extent, by symmetrical baseplates because although their design facilitates all round traverse, in practice large changes in line could necessitate bedding-in again because of the hole left by previous firings in the opposite direction.

Fig. 59.   Mortar with angled baseplate

Baseplate Design

There are no fixed rules for the design of baseplates with the shape, area and
spikes or protrusions on the underside varying with the firing pressures they
have to withstand and the bearing strengths of the soils in which they are to be
used. Many baseplates have holes in them to vent any air that may become
trapped between the baseplate and the soil. These holes prevent trapped air
from providing an elastic cushion when the mortar is fired and also reduce suction that might otherwise exacerbate the task of extracting the baseplate.

Most baseplates have a socket to receive a ball joint at the end of the barrel.
The ball joint and socket are not completely spherical in most modern designs.
Instead the ball has two parallel, flat sides which lock into the socket after being
rotated through 90°. Alternatively, some lighter mortars have their barrels
fixed rigidly to the baseplate. Although this arrangement has the advantage of
simplicity, it inhibits traverse once the baseplate has been bedded-in. Regardless of the method used to connect the baseplate and the barrel, the baseplate
section immediately beneath the axis of the barrel may tend to shear if not
strengthened in some way. This is often achieved by the positioning of a spike
or protrusion beneath the point at which the two components are connected.

Some baseplates have been designed with adjustable surface areas to cater for
varying soil conditions; however, this concept is not without its drawbacks.

Usually the adjustable baseplate comprises a "parent" centre plate with a series of rings that can be added to increase the overall area of the baseplate. The rings must be jointed evenly and firmly to the parent plate. Furthermore, a bending moment is set up at the joint, raising the possibility of distortion that may make disassembly difficult. Moreover, unless the soil conditions are predictable, and in most cases they are not, all sections of the baseplate would need to be carried. If all sections are carried there seems little point in not using the full sized baseplate on all occasions as long as the ground is not exceptionally hard.

Materials

Steel is the most commonly used material in the manufacture of baseplates. Although steel forgings would be preferable, welded mild steel is used extensively. Other materials such as magnesium and aluminium alloys have been used successfully. It is debatable whether it is cost effective to use exotic materials to improve the life of mortar baseplates, because the weight and cost of a spare steel baseplate equates to a small amount of ammunition. Any future improvements to baseplates should, above all, ensure that the result is as cheap and light as possible.

Removal of Baseplates

The quick removal of mortar baseplates after firing is often difficult. If the baseplate has become deeply embedded in the ground the task can become extremely time-consuming, especially if the ground has frozen. In emergencies the time needed to extract the baseplate may be unacceptable which is another good reason for cheap, spare baseplates. The holes provided in some baseplates to prevent a cushion of air forming between the baseplate and the ground also reduce suction that may otherwise increase the time and effort needed to extract the plate. On the other hand the disadvantage of having holes to vent the air is that during firing the soil is forced up through the holes to cover the top of the baseplate which in turn complicates extraction. Various techniques have been tried to help remove baseplates. Besides the obvious expedients such as digging and, in emergencies, the use of crowbars and vehicle-tow ropes, baseplate removal by the use of explosives has also been tried. Explosive removal can be achieved by the use of ballistite cartridges fired from a chamber incorporated in the design of the baseplate. Some mortars such as The Hotchkiss-Brandt rifled mortar also make use of the barrel to lever-out the baseplate.

BARRELS

General Description

Mortar barrels like gun barrels provide the means of imparting direction to the projectile. Except for some light mortars the barrel of a conventional mortar is supported at the required angle of elevation by a bipod, or a tripod, or less frequently by a monopod. The breech end of the barrel is supported by the baseplate.

Compared with gun barrels, for a given calibre, mortar barrels are generally lighter and shorter. A mortar barrel is essentially a tube closed at one end. In most cases they are forged and machined or machined from a solid billet. Some light mortars are constructed from lengths of commercially available metal tubing. There is no need to pre-stress the barrel, as is common practice with modern gun barrels. The external finish is usually smooth although some barrels are finned to provide extra surface area for the dissipation of heat. The usefulness of finning depends largely on the circulation of cool air over the fins. This may not always be possible especially if the weapon is dug-in or vehicle mounted. Although mortar barrels are generally of simpler design than gun barrels they include the means of initiating the charge and, for a conventional, muzzle loading mortar must also provide for "windage".

Windage

The term windage refers to the difference between the diameter of the bore and the diameter of the mortar bomb. If the windage is insufficient, a mortar bomb dropped down the barrel may become trapped by the cushion of air produced in the bore or drop so slowly that its impact on the firing pin or post may not be enough to ignite the charge. Further considerations are that the impact velocity must not be so great that the bomb is damaged, yet it must be sufficient to ensure that the time taken for the bomb to drop down the barrel does not reduce the rate of fire to an unacceptable degree.

The windage dimensions must also allow for the change in barrel dimensions caused by barrel heating and wear. A hot barrel will result in reduced windage and wear will increase windage. The barrel temperature would need to be in excess of about 500°C for a reduction in windage of 0.5 mm for an 81 mm mortar. Nevertheless, when one considers that the minimum windage for a mortar of this calibre is in the order of 0.6 mm, barrel heating can become significant especially if dirt or fouling is present in the bore. Similarly, although wear in mortar barrels is relatively slight the condemnation limit for an 81 mm barrel in British Service is 0.5 mm.

When the charge has been ignited some method of forward obturation must be present. Many different approaches have been tried and perhaps the most common approach in the past has been to provide the bomb with a series of grooves which provide sealing as a result of the baffling effect of the gases moving over the grooves. The main disadvantages of this approach are the turbulence caused by the grooves when the bomb is in flight and the fact that the windage gap to be sealed will vary for the reasons given in the previous paragraph. Although this approach has been used successfully in the past, more recently there has been a trend towards the use of plastic sealing rings that expand under gas pressure to provide forward obturation. Even this approach is not 100% effective, especially at lower charges; however, the results produced can at least be consistent.

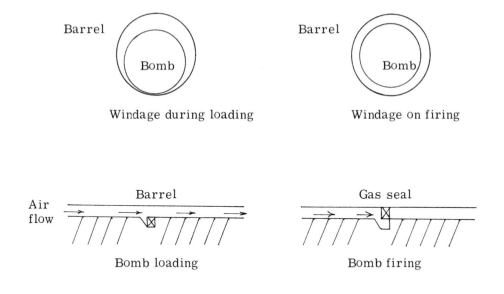

Fig. 60.   Windage

An alternative approach to the problem has been the use of an air valve near the bottom of the barrel. The valve permits the escape of air as the bomb slides down the barrel but is pushed closed by the tail fins as they drop down past the valve seating. Once the charge has been ignited the gas pressure keeps the valve closed until the bomb has left the muzzle and the bore pressure is equalised. The problems with this concept include striking of the valve caused by fouling and the requirement for some degree of clearance between bore and bomb to facilitate loading. Furthermore, the clearance allowed would vary with barrel temperature causing inconsistencies in bore pressures.

Length

The barrel length of an orthodox, muzzle loading mortar is usually restricted by the height at which the loader can safely drop bombs down the barrel at maximum elevation. Clearly the height of the loader and the length of the bomb are two of the major considerations. Ideally the height of the muzzle should not be above eye level. There have been many examples of mortars with barrels longer than 1. 5 metres and unless some provision is made to elevate the position where the loader stands, quick and safe loading drills are not possible. On the other hand longer barrels can produce increased muzzle velocities and hence greater range. Some mortar manufacturers such as Hotchkiss-Brandt (France) and Tampella (Finland) offer a choice of barrel lengths for the same bomb. The 81 mm Tampella mortar barrel is designed in two sections screwed together to improve its manportability.

Strength and Safety

Because mortar barrels do not need to withstand bore pressures as high as guns of equal calibre, their walls are comparatively thin. The British 2 inch mortar had a maximum working pressure of around 1.8 tons/in^2 (27 MPa) while the Israeli Soltam 120 mm mortar has a pressure of 7.0 tons/in^2 (106 MPa). The pressure usually decreases towards the muzzle and in the UK 81 mm mortar it drops to around 1 ton/in^2 (or 15 MPa). The barrel usually accounts for approximately one third of the overall weight of a medium mortar and even less for larger calibres. Efforts to decrease the weight of mortar barrels are probably only worthwhile when manportability requirements are paramount. Provided that the barrel is manportable there seems to be little point in striving for additional small reductions at the expense of degrading the maximum levels of pressure and temperature that the bore can withstand. Moreover, in the manportable role it is the weight of ammunition that is more significant than the total weight of the mortar or the weight of its component parts.

Wear

As mentioned earlier, the wear rates in mortar barrels are low. Abrasive wear is caused by the passage of the bomb's tail fins through the bore. Erosive wear is caused by the action of gases under high pressure, particularly at any point of leakage in the method of obturation used. Of the two abrasive wear is the most significant because erosive wear rates are limited by the low gas pressures in mortar tubes. Wear rates can be reduced by honing the surface of the bore to a smooth finish to eliminate any irregularities in the surface finish together with any surface hairline cracks. Some mortars, such as the Hotchkiss-Brandt 81 mm mortar, have chromium plated bore surfaces; however, this technique hardly seems justified given the cost of the process and the characteristically low wear rates of mortar barrels.

Moisture and Fouling

Mortar barrels are susceptible to the ingress of water during heavy rain because the barrel is always positioned at high angles of elevation when the mortar is deployed for firing. Despite the use of muzzle covers and drills to ensure that the muzzle is never left uncovered needlessly, whenever prolonged periods of firing are required in heavy rain it is difficult to keep moisture out. Even with waterproof charge and ignition components the effect of water ingress on bore pressures and rate of burning can be significant.

Fouling in the bore can be caused by one or more of the following: dirt and grease on the ammunition; traces of lubricants left in the bore; particles of obturating ring deposited on the surface of the bore; and the residue left by any unconsumed portion of the charge. The combined effect of these causes is usually a sticky, black coating in the bore. If this residue is allowed to build up, eventually the windage will be reduced to the extent that the rate of fire will be degraded. Preventative means include careful handling of ammunition to avoid any accumulation

of dirt and grease as well as drills that provide for swabbing out the bore at
regular intervals during firing.

Firing and Trigger Mechanisms

The normal means of igniting the charge is by the use of a fixed firing pin or post
fitted into the closed or breech end of the barrel. In some mortars the pin or
post is screwed into the barrel at right angles to the breech piece. In others it
is screwed in at a small angle (10-20°) to the axis of the bore. The former
approach is usually combined with a removable screw-on breech piece whereas
the latter is often used with solid one-piece barrels. Both systems can work
successfully as long as some provision is made for the removal and inspection
of firing pins or posts. A further consideration for man-packed mortars is the
added weight needed to strengthen the joint of a screw-on breech piece; however
the additional weight is less than about 10% of the total barrel weight and, therefore, would rarely be significant.

Some mortars are fitted with trigger mechanisms so that the bomb can be fired
at any time after the mortar is loaded. In most cases trigger mechanisms are
used because the impact velocity of the bomb at the firing pin is insufficient to
ignite the charge or to enable a final check of lay to be done as is sometimes required with hand-held mortars. An advantage of a trigger mechanism is that it
improves response to calls for fire on pre-arranged targets, as the loading
drills can be done beforehand. Some mortars are equipped with an optional firing
mechanism that can be used whenever the circumstances demand. This approach
is often used in the design of Russian mortars.

BARREL SUPPORTS

The most commonly used form of barrel support is a bipod, although monopods
and tripods are also used. A bipod has a barrel clamp at the top of the bipod
connected to a traversing and elevating gear. Some mortars have shock absorbers and, if fitted, they are usually a part of the bipod assembly. The normal
type of shock absorber consists of one or more cylinders containing springs;
however, hydraulic systems have been used for heavy mortars. Their function
is to assist in returning the barrel into position after firing. A cross levelling
device may also be included in the bipod assembly to ensure that the sight remains
vertical on uneven ground. The legs support the whole bipod assembly and are
usually fitted with plates or shoes near their tips to prevent the bipod from sink-
ing too far into the ground. Sometimes the legs are braced by a cross member
which incorporates mechanical locks to hold them in the firing position. All bipods should allow for easy movement of the legs should there be a need to reposition the mortar beyond the extremes of its limited top traverse. Most bipods
can also be folded to facilitate carriage during redeployment.

As with the other components of mortars, weight is an important consideration
in the design of bipods. There is considerable scope for the use of light alloys
since the bipod is not required to absorb the main firing stresses. Generally
spreaking the bipod needs only to be strong enough to support the weight of the

barrel and withstand rough handling. Nevertheless bipods are usually the most expensive major sub-assembly of a mortar because of the precision needed in the manufacture of traversing and elevating mechanism, shock absorbers, and levelling devices.

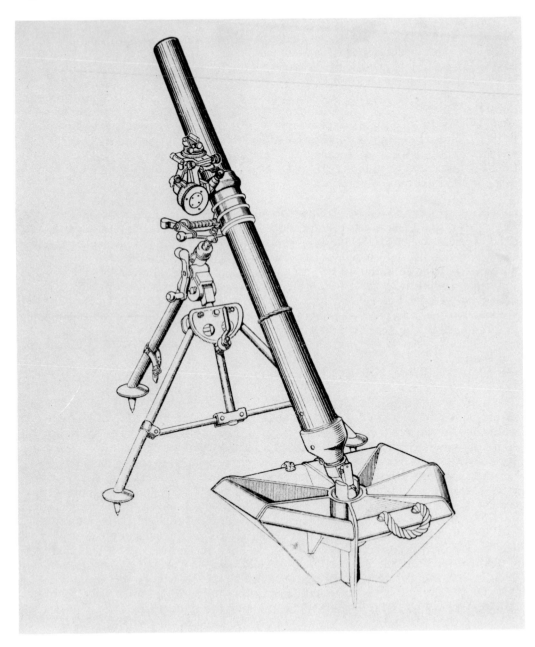

Fig. 61.   Mortar (Tripod)
(British 4.2 inch)

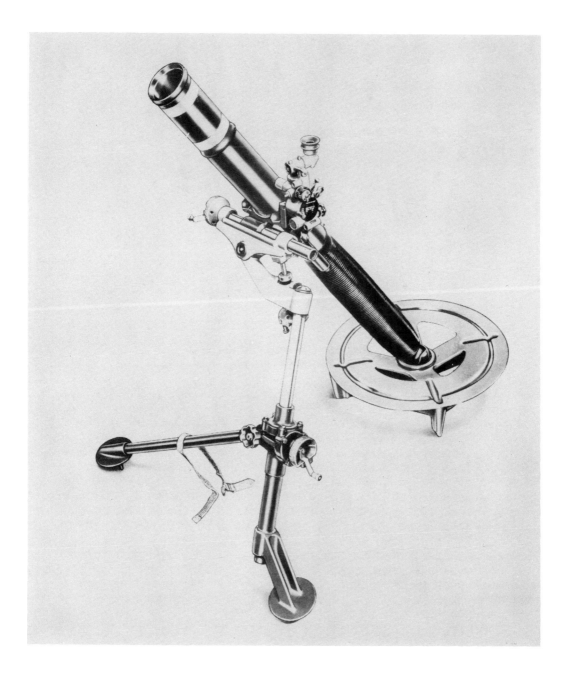

Fig. 62.   Mortar (Bipod)
(British 81 mm)

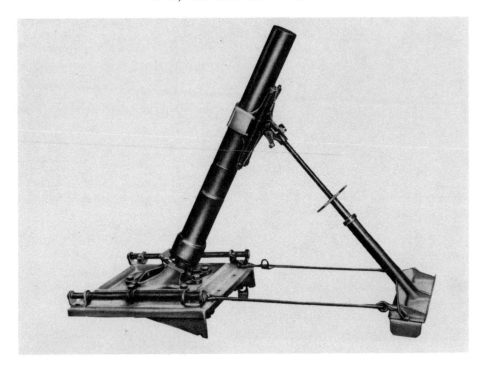

Fig. 63.   Mortar (Monopod)
(American 4.2 inch)

SIGHTS

The more advanced mortar sights are similar in design and operation to the basic form of gun sight described in Chapter 5.   They incorporate the use of fine and coarse scales for both bearing and elevation, as well as some means of cross levelling.   On a normal multi-charge mortar system the range to be achieved is converted into the appropriate elevation needed for the charge in use, the information being extracted from range tables or produced by a small ballistic computer.   Some of the simpler mortar systems permit the use of the actual range units to be set onto the sights.

Most mortars have removable sights so that they can be protected during movement.   When mounted on the weapon the normal position for the sight is in a sight mount attached to the barrel clamp, but wherever it is mounted there must be a link between sight and barrel to maintain the angular relationship between the two in the horizontal and vertical planes.

RIFLED AND BREECH LOADING MORTARS

Rifled

Some of the inherent inaccuracies in mortars can be overcome by the use of rifled

barrels because of the greater stability that can be achieved by spinning the projectile. Furthermore, as discussed in Chapter 3, it is easier to design a spun
projectile for use at supersonic velocities than it is a fin-stabilised projectile.
Rifled mortars can be either breech loading or muzzle loading. In muzzle loaded
versions the mortar bomb's driving band can be pre-engraved and on loading the
engraving on the band must be carefully matched to the rifling in the bore. The
disadvantage of this system is that the maximum rate of fire may be reduced
because of the extra care needed in loading drills. On firing the driving band
distorts to provide forward obturation and to engage firmly in the rifling of the
bore. The Hotchkiss-Brandt 120 mm Rifled Mortar is an example of such a
system.

An alternative approach for a rifled, muzzle loading mortar is to provide the
bomb with a driving band that is not pre-engraved and is mounted so that it does
not exceed the maximum width of the bomb. The bomb then slides easily down
the bore without the driving band engaging the rifling. On firing the band is expanded by a small charge beneath the driving band causing it to engage in the
rifling and provide forward obturation. The disadvantages of this system are the
complexity in bomb design and the likelihood of unreliable obturation because of
uneven expansion of the driving band. The Japanese M 1940 (Type 89) 50 mm mortar is an example of this type of system.

Breech Loading

There have been many examples of breech loading mortars produced and the concept can permit the use of either fin or spin-stabilised bombs. Some of the
earliest examples of modern mortars that appeared during World War I were
breech loading systems. The first German infantry mortar in World War I was
a breech loader and many other countries have since used this design. Such
systems could probably be regarded as unorthodox when compared with typical
modern mortars. Nevertheless, the concept has certain advantages for vehicle
mounted applications and these will be covered later in this chapter. The main
disadvantages of breech loading mortars are access to the breech and the loss of
some potential for high rates of fire. By necessity, breech loading mortars are
fitted with a trigger mechanism to actuate the firing pin: a requirement that sometimes exists for rifled, muzzle loaders too, because the rifling can reduce the
impact velocity of the bomb to the extent that drop firing may become unreliable.

VEHICLE MOUNTED MORTARS

General Description

Mortars mounted on wheeled trailers are currently in service in many armies,
especially those of the Warsaw Pact countries. More recently the increased
emphasis on mechanised warfare has led to greater use of mortars mounted in
tracked vehicles. Trailer mounted mortars towed by light wheeled vehicles have
been used successfully for some time, however, this combination often lacks the

degree of cross-country mobility needed to keep pace with armoured and mechanised infantry units, especially in marginal going conditions.

So far the approach to providing tracked mortar systems has been to take an existing ground-mounted mortar and adapt it to an in-service tracked vehicle chassis:
usually an armoured personnel carrier chassis. In most cases the marriage of
the two works adequately. Examples of such systems are the Tampella 120 mm
Mortar mounted in an American M113 and the UK 81 mm Mortar mounted in
FV432. In addition to their greater mobility, tracked mounted mortars also have
the advantages of extra crew protection, no requirement for bedding-in, and less
time used in preparation for firing. The main disadvantage is that a degree of
strategic and tactical mobility is lost, particularly in terms of airportability. It
follows, therefore, that most armies will continue to have a requirement for
ground-mounted systems.

The design considerations for tracked mortar systems are best considered under
two general headings. Firstly, the systems that can be regarded as hybrid
systems similar to those that have already been described; secondly, purpose-
built tracked mortar systems or systems that are produced as special variants
of a family of vehicles. The latter type of system would have the vehicle compartment specially fitted for its primary role as a mortar carrier and the mortar
itself may be quite different to ground-mounted versions firing the same ammunition. Both types will now be discussed.

Fig. 64.   French AMX 13 120 mm Mortar Carrier

Note: 1.   This is an example of a hybrid solution. It is a
            120 mm mortar adapted to an AMX 13 chassis.

      2.   Note the baseplate fitted to the front of the vehicle
            for the ground role.

Hybrid Systems

There are some obvious advantages in a hybrid system, the main one being that it is the cheapest solution. Not only can an existing in-service chassis be adapted for the task, but obsolete or obsolescent tanks or armoured personnel carriers can be modified to take mortars. The disadvantages are that the mortar barrel may be unnecessarily light for its vehicle-borne role: its length may not suit the vehicle compartment; and some alternative arrangement will be needed to replace the baseplate.

Barrel weight is often restricted in ground-mounted systems by the need to ensure that the mortar, or at least its main component parts, remain manportable. This constraint is usually superfluous for a tracked system with the result that the mortar's range and ability to dissipate heat can be less than the achievable maximum for the calibre. The barrel length of a ground-mounted system can pose problems by being either too long or too short for the vehicle compartment. If the barrel is too long it can restrict traverse. If it is too short, blast over-pressures at the muzzle can be extremely hazardous in the confines of the vehicle compartment. The arrangements made for adapting a ground-mounted system to a vehicle so that recoil is absorbed has resulted in a variety of concepts. In some cases rigid mountings are used, with the breech socket supported by a steel plate in the vehicle floor which in turn may need to be strengthened to withstand the force of recoil. The Israelis have mounted a 120 mm Tampella Mortar on a half-track chassis using this approach. Alternatively, the barrel can be mounted on a central plinth or on a turntable adaptor connected to the floor by a flexible mounting. An example of the 81 mm Adaptor for FV432 is shown in Fig. 65.

Fig. 65.   Adaptor for UK 81 mm Mortar in FV432

Although a tracked vehicle chassis is usually rigid enough to withstand firing
stresses, albeit with minor modification in some cases, the vehicle's suspension
system is a further consideration in the overall effect of recoil. It may be necessary to lock-out the suspension system in much the same way as is done with
some SP guns to prevent damage to the suspension. Alternatively the suspension
system could be used to help cushion the effects of firing.

In a hybrid solution, elevation and traverse limits can sometimes be a problem.
Typically, a barrel designed for a ground mounting is about 15 calibres in length.
If the barrel is too short to clear the sides of the vehicle, or the hatch if one is
used, the barrel will have to be mounted on an adaptor to raise the muzzle to a
safe height. The result may be that the loader cannot reach the muzzle without
some form of stand or footrest being provided, all of which encroaches on the
available space within the vehicle. Traverse can often be limited by vehicle
hatches, antennae and barrel length. Some systems permit limited traverse and
require that the vehicle tracks be moved if targets are to be engaged beyond top
traverse.

Purpose Built Vehicle-Mounted Systems

Many, if not all, the inherent problems of hybrid systems, could be overcome by
designing the mortar and its carrier specifically for the role. The mortar could
be mounted in a number of ways to improve the degree of protection afforded to
the crew. In existing hybrid systems the level of overhead protection provided
to the crew when firing is negligible or at best limited. It is difficult to overcome
this problem with a muzzle loading mortar; however, a breech loaded mortar
would enable the design of a system in which the crew would be given overhead
protection with the mortar mounted in a turret for 6400 mils traverse. The savings in space within the vehicle could be considerable too, thus providing an increased ammunition carrying capacity. Space in existing hybrid systems is at a
premium and in general this constraint, rather than weight considerations, tends
to dictate the number of mortar bombs carried. Despite the advantages of a
purpose-built system most armies obviously view such systems as too expensive
and perhaps too complex. The cost and complexity may begin to approach that of
an SP gun but without its advantages. Nevertheless, the increased use of vehicle-
mounted systems could mean that they will, in future, be designed more delibera-
tely than in the past.

Barrel Heating

The problem of barrel heating can be compounded by mounting it in a vehicle.
The use of a finned mortar barrel to help dissipate heat is far less effective when
the weapon is sheltered from any prevailing breeze. Figure 67 shows some
typical comparative figures for heat dissipation in smooth and finned barrels.
The improvement gained in the regard by a finned barrel when a 10 knot breeze
is blowing increases by a factor of almost 2. Much of this additional advantage
could be lost if the mortar is vehicle mounted. On the other hand, if the barrel
is purpose-built for a vehicle mounted role it could possibly be of heavier construction and thus provide a greater heat sink which would alleviate the barrel

heating problem. More complex arrangements for barrel cooling such as water cooling and forced air cooling are also possibilities in vehicle mounted systems; however, it is unlikely that this degree of sophistication would ever be accepted for a hybrid solution.

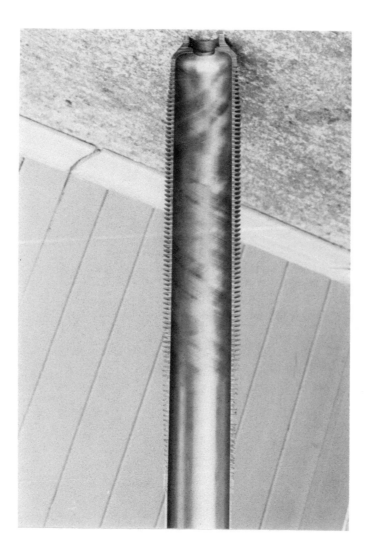

Fig. 66.   Finned mortar barrel

Note: Example shown is a cross section of a UK 81 mm
mortar barrel.

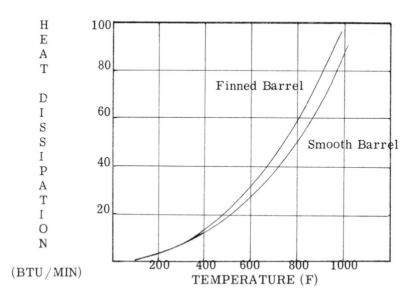

Fig. 67.   Heat dissipation in mortar barrels

Note: Photograph above shows an 81 mm barrel partially
finned but with the top section smooth.

HAND-HELD MORTARS

A large number of light, hand-held mortars have been employed and are still being produced. Their main purpose is for use by infantry sub-units to provide short range, indirect fire. Most of these weapons are not equipped with sights and the firer simply judges the line to the target and lays the barrel along it. The effectiveness of this system depends on the size of the area to be engaged but it can be sufficiently accurate for the maximum range that these mortars can achieve, which is in the order of 5-700 metres. Range is altered by changing the elevation of the barrel by hand. Sometimes a sling is provided for the firer to place his foot in, the sling being marked at the appropriate length above which the muzzle must be positioned to achieve a given range. The baseplate, if one is used, is usually small and may be manufactured by pressing rather than forging or casting as is the case with larger mortars.

Fig. 68.   Hand-held mortar (firing)

The mortar shown in the figure above is a British 51 mm man-portable mortar. It is representative of the state-of-the-art in modern light mortars. The total weight of the mortar is a little over 2 kilograms and it can be carried easily on a sling by one man as shown in Fig. 69. It can fire high explosive, smoke and illuminating bombs to a maximum range of 750 metres, with a maximum rate of fire of eight bombs per minute for two minutes. It is not possible inadvertently to double load the mortar because the second bomb will protrude from the muzzle.

This is a safety feature that is incorporated in the design of several other mortars. The ranges achieved by the fixed charge ammunition can be varied by the use of a barrel insert that causes a drop in barrel pressure and hence a lower trajectory for a given elevation. Low angle or even direct fire is possible, with the limitation on minimum range determined by the arming time of the fuze.

SUMMARY

Mortars provide a simple, flexible, cheap means of providing indirect fire. The introduction of expensive materials and complex design reduces these inherent advantages.

Fig. 69.   Hand-held mortar (carried)

SELF TEST QUESTIONS

QUESTION 1   List the main characteristics of a conventional mortar that
              distinguish it from a gun.

              Answer   ....

                        ....

QUESTION 2   Name the main components of a mortar.

              Answer   ....

                        ....

                        ....

                        ....

QUESTION 3   Explain the term "bedding-in".

              Answer   ....

                        ....

QUESTION 4   Why are some baseplates manufactured with holes in them?

              Answer   ....

                        ....

QUESTION 5   Explain the term "windage".

              Answer   ....

                        ....

                        ....

QUESTION 6   Some mortar barrels are finned. Explain the reason for this
              and the advantages that may accrue from finning.

              Answer   ....

                        ....

                        ....

                        ....

QUESTION 7   What is the main limitation on the length of a muzzle loading
                  mortar?

                  Answer   ....

QUESTION 8   List two manufacturing techniques that are used in the
                  manufacture of mortar barrels to help reduce wear.

                  Answer   ....

                           ....

QUESTION 9   What advantages are there in a mortar with a trigger
                  mechanism instead of a fixed firing pin or post?

                  Answer   ....

                           ....

                           ....

                           ....

QUESTION 10   List the disadvantages of rifled mortars and breech loading
                  mortars.

                  Answer   ....

                           ....

                           ....

                           ....

                           ....

QUESTION 11   Vehicle mounted mortars are usually a combination of an
                  existing vehicle chassis and an existing ground mounted
                  mortar. Explain the reason for this and state whether you
                  think it is the best approach and why.

                  Answer   ....

                           ....

                           ....

                           ....

                           ....

                           ....

QUESTION 12   What are the main features of hand-held mortars and what
are their limitations?

Answer   . . . .

. . . .

. . . .

. . . .

. . . .

. . . .

ANSWERS ON PAGE 196

7.

Free Flight Rockets

GENERAL PRINCIPLES

If a gas is held under compression in a closed tube the pressure is equal and opposite in all directions (see Fig. 70). If an opening is present at one end of the tube, as in a rocket, and the pressure maintained by burning propellent, the pressure at the closed end is greater than at the open end. The compression energy represented by the dotted lines in Fig. 71 is expended in giving velocity to the escaping gases. The effect of the escaping gases is to cause the rocket to move in the direction of the closed end; however, because the mass of the escaping gases is less than the mass of the rocket, the rocket will move in the opposite direction at a slower velocity.

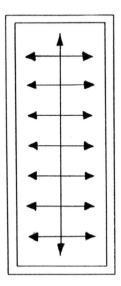

Fig. 70.　Gas pressure in a closed tube

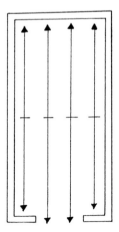

Fig. 71.   Gas pressure in an open tube

The principle that governs the action of a rocket in the circumstances just described is known as the "conservation of linear momentum":

Linear momentum = mass x velocity

The acceleration of the rocket can be calculated as follows:

$$\text{Acceleration} \ = \ \frac{\text{mass of fuel consumed per second}}{\text{mass of rocket}} \ \text{x} \ \ \text{velocity of gases}$$

Acceleration is directly proportional to the rate at which fuel is consumed.
Further, if the fuel can be burnt at a constant rate the acceleration will increase
as the fuel is used. The greater the velocity of the escaping gases the greater the
velocity attained by the rocket. The velocity of the escaping gas depends on the
fuel used, the pressure at which the fuel is burnt and the "venting conditions", or
the conditions under which the gases are expelled from the open end of the rocket.
Of these the fuel used is perhaps the most important, although all the factors mentioned are important. In general, liquid fuels produce higher energy values and
better performance, but there are problems associated with the application of
liquid fuels to free flight rockets (FFR). These will be covered later in this
chapter. The formula for calculating the maximum velocity attained by a rocket
is as follows:

$$\frac{\text{Maximum velocity}}{\text{Velocity of gases relative to rocket}} \ = \ \log_e \frac{\text{Initial mass of rocket}}{\text{Mass of rocket when fuel consumed}}$$

The velocity calculated by the above formula must be corrected for the effects of
gravity and air resistance. The maximum velocity achieved does not depend on
the rate at which the fuel is consumed and the rocket will reach the same maximum velocity regardless of whether it has a high or low acceleration. The important parameters for maximum velocity are the velocity of the escaping gases
and the overall proportion by weight of fuel in the rocket. Two other general

characteristics of rockets are worthy of mention. The first is that the propulsive force developed is unaffected by the velocity of the rocket. The second is that the thrust developed is independent of the atmosphere.

The main components of an FFR are the motor (including the combustion chamber and the nozzle) and the warhead (including the fuze).

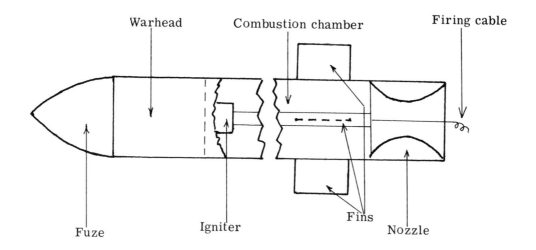

Fig. 72.   Main components of a free flight rocket

These components will now be described together with the important considerations in the design of rockets and their launchers. Finally a couple of examples of modern FFR systems will be given to illustrate the present state-of-the-art.

ROCKET MOTORS

Casing

A rocket motor for an FFR is simply a casing to provide a combustion chamber in which the charge can burn. Its forward end is closed and attached to the warhead. It contains an igniter and the rear end has a nozzle. The motor casing must be strong enough to withstand the high temperature and pressures reached during combustion. If the casing is susceptible to bending when the motor is fired or if it is not geometrically precise there can be problems in dispersion at the target for reasons that will be covered later. Cold worked flow forming manufacturing techniques have been used for rocket motor casings and these have produced good results in terms of material strength and the precision in reproducing the dimensions required. Other manufacturing techniques include the use of glass fibre reinforced plastic, deep drawn steel tubes, helically wound steel strips, and steel sheets wrapped and welded to form a cylinder. FFRs use boost motors which are motors that produce high levels of thrust over a short period of time. Sustainer motors producing relatively low thrust over a longer period of time are not used for FFRs.

Propellents and Igniters

Solid propellents are generally preferred to liquid propellents for FFR motors, although it is possible to use liquid propellents in an FFR. Indeed liquid propellents have certain advantages in terms of energy values and performance. The main disadvantages in their use is the complexity in design of the rocket motor together with the associated cost. Although solid propellents have lower energy values and are heavier for a given performance, they are preferred for FFR applications because of their reliability and simplicity in use. Liquid propellents are, however, widely used for long range guided weapons where their high energy values, long burning duration and adaptability to intermittent controlled operation becomes important.

An igniter is essential for a solid propellent rocket and usually it is initiated electrically. The igniter must be positioned so that it causes the charge to commence burning on all the available burning surface. For this reason the igniter is usually large and is fitted at the forward end of the charge so that as the flash from the igniter moves rearwards towards the nozzle it passes over the exposed surfaces of the propellent.

Nozzles

The normal type of nozzle used in an FFR is a convergent-divergent nozzle, sometimes called a "De Laval nozzle". The purpose of the nozzle is to change heat and pressure energy into kinetic energy. The shape of the nozzle achieves this by decreasing the cross sectional area of the opening through which the gases can escape. Because the mass flow of gases through the nozzle remains constant the flow is accelerated. As the gases emerge from the narrow section or throat of the nozzle they are expanded to low temperature and pressure, reaching a high velocity. The thrust developed by the gases is produced by the rate of change of momentum in passing through the nozzle with the result that a propulsive force is exerted on the rocket.

Figure 73 shows the critical dimensions of the nozzle. The inner slope (A) produces a smooth, non-turbulent flow of gas. The throat of the nozzle must allow sufficient free space for the gases without throttling but at the same time prevent a too rapid escape so that pressure can be maintained within the rocket. The outer slope (B) uses the lateral expansion of gases to obtain additional forward thrust. The angle of the outer slope is usually about 30 degrees. The nozzle must be able to withstand the erosive action of hot, compressed gases moving at high velocities while maintaining its internal geometry throughout the operation. The effect of the escaping gases moving at high speed is most critical at the throat of the nozzle. The material used for the nozzle should have a high melting point and good thermal conductivity. Additionally, it should be robust enough to resist the abrasive action of the gases. Metal oxides or carbides and asbestos components are some of the materials that can be used. When it is desirable to use a material which is expensive, in short supply, or can withstand only a limited mechanical stress, it is sometimes made in the form of an insert which is fitted into a recess bored out to receive it either as a shrink fit or a force fit. In some cases the nozzle may be screwed on so that it can be removed to facilitate the inspection of the charge.

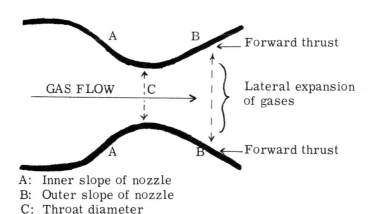

A: Inner slope of nozzle
B: Outer slope of nozzle
C: Throat diameter

Fig. 73.   Critical dimensions of a rocket nozzle

WARHEADS

FFRs can be adapted to carry a wide variety of warheads including nuclear (calibre 150 mm and above), HE, chemical, pre-formed fragments, as well as sub-munitions including terminally guided sub-munitions. They also have a greater degree of inherent flexibility than guns. A general feature of fin-stabilised rocket ammunition has been the ability to fit a different type, weight and size of warhead to the same motor. Spin-stabilised rockets are more limited because any alteration in warhead shape and weight may upset the stability of the rocket. Fin-stabilised rockets and rockets stabilised by both fin and spin have warhead limitations too; however, the limits within which the same launcher and motor can be used are much greater. Moreover, with some FFRs the fins are larger than required for stability purposes and changes in warhead can be effec-ted with little or no diff erence in stability. Some launcher tubes with helical guide rails have been designed so that the warhead protrudes from the tube; therefore the diameter of the tube does not limit the size of the warhead. Nowa-days the trend is towards larger calibre rockets firing to long ranges with the warhead flexibility being more a function of the rockets ability to carry either a full calibre warhead or sub-munitions.

Projectiles fired from guns need comparatively thick walls and bases to withstand the high acceleration stresses in the bore. For this reason the proportion of

filling to total projectile weight is usually well below the optimum. Gun projectiles need to withstand accelerations as high as 20,000 g, whereas with FFRs the
figure is in the order of 30-50 g.

The fuze is normally positioned inside the ballistic nose cone of the rocket. Conventional HE warheads are usually designed to function on impact and time fuzes
are fitted to rockets carrying payloads of sub-munitions. In modern rocket
systems such as the French RAFALE 145 mm Multiple Rocket Launcher and the
American Multiple Launch Rocket System (MLRS) the fuzes incorporate solid
state electronics instead of the conventional mechanical fuze mechanisms previously fitted to rockets and still used for most conventional gun systems. It is
also possible to set, remotely, electronic fuzes of this type which is obviously an
important factor in decreasing response times, especially when high rates of fire
are needed.

Rockets are ideally suited as carriers for terminally guided sub-munitions such
as SADARM (Sense And Destroy ARMour). A number of sub-munitions of this
type can be carried in a single rocket from the MLRS system to a maximum range
of approximately 30 km. When the rocket reaches the target area the SADARM
sub-munitions are ejected and descend by parachute at about 9 m/sec. As they
descend each of the cylindrical sub-munitions rotates at 3-4 revolutions per
second while its sensor scans the area below. The sensor detects targets within
its area of authority, or limits of its scan and manoeuvrability, and calculate the
precise moment for firing a self-forging fragment downwards onto the target.
Unlike the US COPPERHEAD cannon launched guided projectile, discussed in
Chapter 8, no external designator is needed. Terminally guided sub-munitions
such as SADARM delivered at long ranges by multiple launch rocket systems
obviously have great potential for depth fire applications, especially counter
battery and anti-armour tasks.

            LAUNCHERS

A rocket launcher serves to support and aim the rocket. A launcher in its simplest form may be expendable; however, most modern launchers are designed for
reloading. Rocket launchers may be built to carry a single rocket as in Fig. 74
below, or a number of rockets. Since a rocket moves forward by ejecting gases
backwards there is no significant recoil on the mounting apart from a small
amount of friction between the rocket and its guide rails. If there is to be no
recoil, the escaping gases must be allowed to pass unimpeded to the rear of the
launcher. In practice, however, this is difficult to achieve because the escaping
gases, in expanding, will impinge on parts of the launcher even though the net
effect is small. Unlike a gun there is no trunnion pull to consider; therefore an
increase in maximum range does not necessarily mean an increase in the weight
of the launcher as may often be the case with a gun.

Because there are no great recoil forces to withstand, the mounting for a rocket
system need only be as heavy and as strong as is required to support and in some
cases transport the desired size, weight and number of rockets. It is possible,
therefore, for an armoured SP rocket launcher such as MLRS to transport and
fire 12 rockets each weighing more than 270 kg.

Fig. 74.   Honest John Rocket

There are two main types of launcher: positive length or "rail launchers", and
"zero length launchers". A zero length launcher is one in which the first motion
of the rocket removes it from the restraint of the launcher. The purpose of a
zero length launcher therefore is simply to hold and point the rocket in the required direction. It does not influence the subsequent trajectory of the rocket.
Although zero length launchers can be lighter and smaller than a rail launcher,
they are generally unsuited for FFR system applications because their use results in high initial dispersion. They are, however, normally employed for
guided weapon systems.

In contrast, a positive length rail launcher is long enough to influence the flight
of the rocket after it has begun to accelerate. The term "rail launcher" is used
to describe a wide variety of launchers including tubes and ramps as well as
rails. Tubes seem to be favoured in modern FFR systems because the tube can
be used to provide protection from shell splinters and small arms fire. In addition, tubes can provide good support for the rocket as it accelerates on the
launcher and can be readily adapted to impart spin. The length of rail used is
often a compromise between a number of conflicting requirements such as: the
dimensions of the vehicle on which it is mounted; the weight and length restrictions imposed by the tactical employment; and the minimum length needed to
ensure that the rocket is projected in the direction of the rail/s. In some cases
it is impractical to support a rail along its entire length; therefore, if the rail is
long it can be more susceptible to deflection as the rocket moves along it. The
ratio of the time of rail travel to the burning time of the rocket motor is the key

to the effectiveness of the length of the launcher rail in reducing dispersion. The
time of travel on the rail can be expressed as follows:

$$t \; = \; \sqrt{\frac{2s}{a}}$$

Where t   =   time
        s   =   rail length
        a   =   the acceleration of the rocket

Clearly a rocket motor with a short burning time producing a high acceleration
will be an advantage in this regard. If this requirement can be met, a relatively
short rail will suffice.

STABILISATION

The advantage of spin stabilisation is the degree of consistency that can be
achieved in the dispersion of rockets at the target. Spin stabilisation was used
for most FFRs in World War II. The disadvantage is that as rockets have a high
length:diameter ratio, spin stabilisation is often difficult to achieve because the
rate of spin needed increases with the length:diameter ratio. The analogy sometimes used is that of a spinning top: a long, narrow top is more difficult to keep
upright when spinning than a short, broad one. Consequently, the length of a
spin-stabilised rocket is limited to approximately 6 times its calibre if stability
in flight is to be maintained. This limitation makes it impossible to achieve as
good a ratio of rocket weight to frontal area as with a long, fin-stabilised rocket.
Examples of spin-stabilised FFRs include the Soviet BM24 240 mm multi-rail
launcher and the World War II German 150 mm Wurfgranate rocket. Spin can be
imparted by the launcher or by using a number of nozzles set at an angle to the
longitudinal axis of the rocket instead of a single nozzle. In the latter method
the escape of gases to the rear produces a "catherine wheel" effect, with the
spin conveniently increasing in proportion to the speed of the rocket. The
Wurfgranate rocket employed this technique.

The alternatives to spin stabilisation are fin stabilisation or a combination of
both fin and spin stabilisation. The latter alternative is most commonly used,
with a slow rate of spin or rotation being imparted to a fin-stabilised rocket
either while it is still on the launcher or by auxiliary motors that are activated
after launch. The former method is commonly used in modern FFR systems: the
latter was used in HONEST JOHN. The size of fins on a rocket is usually a compromise between the need for large fins to reduce errors caused by thrust misalignment as well as errors at the moment of launch and small fins to reduce
errors caused by surface cross winds. These errors will be discussed later in
this chapter.

When rotation or spin is combined with fins, the spin can help offset the effects of
thrust misalignment as well as any asymmetric forces produced by fin misalignment. If left uncorrected, both thrust misalignment and fin misalignment can
produce an undesirable degree of dispersion. The exact position of the fins in

relation to the nozzle can also be critical. If they are fitted too close to the nozzle the gaseous plume of the motor efflux may degrade their efficiency.

ACCURACY AND RANGE COVERAGE

Thrust Misalignment

When a projectile leaves the muzzle of a conventional gun and has travelled beyond the influence of gases emerging from the muzzle, the forces that affect the shape of its trajectory are wind, air resistance and gravity. Assuming the projectile is spinning at a rate that ensures stability, wind and drift are the only factors that will cause it to deviate from its initial line, both of which can be calculated and allowed for before firing. With a rocket the difference is that thrust is applied by its motor after it has left its launcher. If the direction of thrust does not pass through the centre of gravity of the rocket a condition called "thrust misalignment" occurs, which results in a proportionally large error in line at the target unless the rocket is spun.

If rockets could be manufactured perfectly with the thrust accurately aligned to pass through the centre of gravity their dispersion at the target could be reduced. In practice, however, this is difficult to achieve. It is not simply a matter of positioning the propellent centrally. The nozzle exit cone axis and the thrust line of the emergent jet must also be aligned with the centre of gravity. In addition, unless the rocket casing or tube is perfectly symmetrical it may tend to bend under the high internal pressures generated by the rocket motor, causing mechanical tolerances to be emphasised. Bending can also be induced by asymmetric heating from the rocket motor. If thrust misalignment is present the rocket tends to rotate and yaw (see Fig. 75). Careful design and manufacture to close tolerances is obviously important in reducing thrust misalignment but this increases the cost of the rocket. Other methods of minimising the effects of

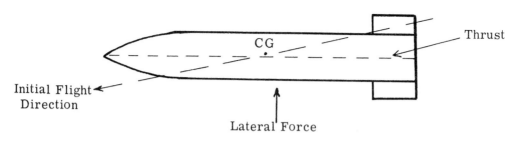

Note: Rocket will rotate clockwise
CG = centre of gravity

Fig. 75.   Thrust misalignment

thrust misalignment are to impart a small amount of spin to the rocket, either by
the launcher or after launch, and to reduce the motor burning time. If the motor
burning time can be made very short, the dispersion caused by thrust misalignment is decreased. Motor burning times of about 1-3 seconds are common and
even shorter burning times are desirable. In general, the shorter the burning
time the higher the pressures in the motor which in turn demands a heavier and
more robust rocket casing. A short burning time is also useful in reducing surface cross wind effects.

Surface Cross Wind Effects

The effect of surface cross winds on the stabilising fins of a rocket can be large.
The effect is to turn the nose of the rocket into the wind (see Fig. 76). The
lower the velocity of the rocket the greater the influence of a cross wind, hence
the effect is greatest immediately after launch. As the velocity of the rocket increases the head wind induced by the movement of the rocket itself tends to
straighten the rocket; however, the trajectory may still be offset at an angle to
the original line of fire. The larger the fins in relation to the other external
dimensions of the rocket the greater the influence of surface cross winds. Thus
a compromise must be reached between a fin size large enough for stability, yet
not so large as to magnify the effect of surface cross winds. If a fin-stabilised
rocket also has an amount of spin to help limit dispersion, the fins need not be as
large and therefore the effect of surface cross winds will not be as marked. The
shorter the burning time of a rocket motor the greater the acceleration and hence
the smaller the surface cross wind effects.

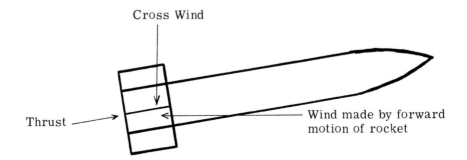

Note: Cross wind from left makes rocket deviate
        to left. Head wind will cause nose of rocket
        to dip.

Fig. 76.   Effect of surface cross wind on finned rocket

Spin-stabilised rockets without fins behave differently in surface cross winds.
The resultant of the forces caused by cross winds acts nearer the nose of the
rocket than the tail. Consequently a cross wind from left to right will cause a
deviation of the trajectory to the left. The magnitude of the deviation will vary

with a number of things, including the rocket's velocity and shape. Nevertheless its magnitude is generally similar to or less than that for a rocket using fin and spin stabilisation.

Velocity at All Burnt

The range dispersion of rockets is affected by changes in the rate of burning of the propellent. Even a small variation in the rate of burning can make a signifi- cant difference in the velocity of a rocket at all burnt. In a typical solid fuel FFR the propellent is shaped and fitted into the rocket in such a way that burning takes place simultaneously over a relatively large area of the propellent. Propellent shapes that present a large available area, such as hollow cylinder and multiple grain hollow cylinder, are frequently used. In addition, the weight of the propel- ling charge is adjusted for each lot of propellent to give uniform performance. The burning rate of a charge is affected by temperature and pressure: the higher the temperature or pressure the faster the burning rate. In the extreme, if a rocket is fired at temperatures higher than those for which it is designed, the pressure can build up faster than the nozzle can release it which will result in increased dispersion or may even cause the rocket casing to rupture. Similarly, if the rocket is fired at temperatures below the specified limit the charge burns slowly, dispersion is increased or the rocket could range erratically short. One method of achieving a more constant rate of burning is to add a small percentage of lead salts (about 2%) such as lead aspirate or lead stearate to the propellent. The use of such additives is called "platonisation" and the result is a more cons- tant rate of burning within specified limits of pressure.

Launcher Induced Errors

An important source of inaccuracy is instability during the firing of a salvo or ripple. Even though rockets are said to be recoilless, there can be sufficient movement during firing to disturb the lay of the system resulting in increased dispersion. The problem can be exacerbated when the launcher is vehicle moun- ted because of movement in the running gear or in the suspension of the vehicle. One method of overcoming this problem is to provide jacks and/or spades for the launcher vehicle; however, the weight penalty may often be unacceptable. An alternative approach is to use hydraulic shock absorbers that return to their original position, but their reaction may not always be fast enough to trim the vehicle before the next rocket is launched. More recently there has been a trend towards the possible use of automatic control systems that incorporate inertial sensors to measure and correct for firing loads. Although the technology is available to make the use of such systems feasible, the cost might only be justi- fied for larger calibre, long-range systems.

A further problem often associated with launchers is the difficulty in ensuring that the shoes or lugs holding the rocket to its rails release simultaneously. If they do not a "tip-off" effect occurs. Tip-off is the tendency of the rocket to tip downwards at launch. It occurs when the forward part of the rocket is unsuppor- ted having cleared the rail or tube, while the rear section is still being held on the rail. The problem seems to have been solved in most modern launcher systems.

Range Coverage

The accuracy of FFR systems has improved greatly since World War II with system accuracies in the order of 1% of range or better now being attainable. Nevertheless they still cannot compete with conventional gun systems for many tasks, particularly close support tasks, because of their inferior range coverage and reloading times. "Spoilers" or air brakes can be used to increase the drag on a rocket in flight and thus alter the trajectory and the range achieved at a given quadrant elevation. For example, the French RAFALE 145 mm rocket system has air brakes positioned between the tail fins at the rear of the rocket. If required, the air brakes are actuated at the launcher before firing and they deploy at the same time as the folding fins deploy on leaving the launcher. Their effect is to reduce the minimum range of the rocket from 18 km to 10 km.

It is possible to design a rocket with sets of air brakes that present different surface areas to vary the range. Nevertheless the use of air brakes can be regarded, primarily, as a means of reducing minimum range rather than a method of achieving range coverage similar to a conventional multi-charge gun/howitzer. Mechanical loading systems have slashed FFR reloading times too. For example, the claim for the Italian FIROS 25 122 mm rocket system is 5 minutes to reload 40 rockets. Regardless of the advances made in reloading times it is highly unlikely that FFR systems will ever be able to match conventional gun systems or mortars at sustained fire rates.

EXAMPLES OF MODERN FFR SYSTEMS

In response to the Warsaw Pact's rocket systems as typified by the BM21 multiple rocket launcher, three of the NATO allies (UK, FRG and Italy) collaborated in an FFR project for a weapon system called RS80. The RS80 was to have a range of 40-60 km depending on the warhead used and was designed to complement conventional 155 mm weapon systems. After the RS80 project was discontinued, the UK and FRG turned their attention to the US Army's MLRS produced by the Vought Corporation while the Italian firm SNIA Viscosa produced two alternative systems: the FIROS 6 51 mm system and the FIROS 25 122 mm system. Although these three systems are by no means the only ones produced in the Western world, MLRS and FIROS 6 are good examples of the state of the art at the opposite ends of the performance spectrum.

MLRS consists of two pods (each containing six rockets) mounted on a tracked chassis (see Fig. 77). The rocket pods are housed in an armoured container that provides a level of protection equal to that elsewhere on the vehicle. Each rocket is in a fibreglass launch tube. On launch the caps sealing the tubes are blown off. Helical rails in the tube rotate the rocket at about 11 cycles per second. Each rocket is fitted with discarding sabots for the purpose of guiding the rocket along the helical rails. Once the rocket has left the tube, 4 spring loaded fins are deployed just forward of the nozzle.

Fig. 77.   Multiple Launch Rocket System (MLRS)

Like all multiple launch systems, one of the main features of MLRS is its ability
to deliver a large number of projectiles in a short period.   The system's 12 x
227 mm rockets, each weighing more than 270 kg, can be fired in a ripple in
about a minute.   If necessary, one man can load and fire the system, although a
three-man crew is needed for sustained operations.   The launcher vehicle is provided with an electrically-operated boom and winch arrangement for reloading.
Although the loading arrangements can be controlled by one man and completed
in a few minutes, the MLRS system is not capable of providing sustained fire
over a long period in the manner that gun systems can.   Although this can be regarded as a good argument for the retention of gun systems, it is not really a
weakness of MLRS because MLRS will be employed on a "shoot and scoot" basis
for reasons of protection.   To facilitate rapid redeployment and engagement of
targets, each MLRS is equipped with its own fire control system.

The onboard navigation system will enable MLRS to compute the coordinates of
the system before firing.   In this way the MLRS launchers will be independent
of pre-surveyed sites and the total time out of action during redeployment will be
kept to a minimum.

The MLRS system weighs more than 22,000 kg when fully loaded which limits its airportability to large aircraft such as the C-141 Starlifter. Its tactical mobility is excellent having a maximum speed in excess of 60 km/hr, the ability to accelerate from 0 to 48 km/hr in less than 20 seconds and the capacity to negotiate rough terrain to an extent similar to that of other modern tracked vehicles.

The Italian FIROS 6 is an example of a small free flight rocket system. It consists of a 48-tube launcher that fires 51 mm rockets with a 2.2 kg warhead, the total weight of each rocket being 4.8 kg. The rockets can be fired either singly or in a ripple at a rate of 10 rockets per second to a maximum range of about 6.5 kms. The FIROS 6 system can be mounted on a small 4 x 4 vehicle such as a Land Rover or the FIAT 6614 wheeled APC (see Fig. 78). Although the warhead is small, the mobility and high rate of fire of the system could make it well suited to operations in difficult terrain, especially where airportability is important.

Fig. 78.   Italian FIROS 6 Multiple Launch Rocket System

SUMMARY

FFRs provide a means of delivering massive firepower, in a short time, at long range and from comparatively light equipments. Furthermore, they are ideally suited to the new range of improved sub-munitions being developed. Despite their logistic penalties and the ease with which they can be detected, we shall see more FFR systems in Western Armies, probably in place of heavy guns.

APPENDIX 1 TO CHAPTER 7

COMPARISON OF MORTARS, GUNS AND FFR

Equipment (kg)	Weight (kg)	Projectile Weight (kg)	Max Range (m)	Muzzle Velocity (m/sec)	50% Zone at $\frac{2}{3}$ Max Range (m)
81 mm Mor (UK)	40	4.3	5650	249	30 x 60
120 mm Mor M-40 (Finland)	285	13.3	6400	317	34 x 50
105 mm Towed How M2A2 (USA)	2258	14.97	11000	472	6 x 50
105 mm SP Gun ABBOT (UK)	16800	16.0	17300	708	7 x 64
110 mm SP Rocket System (FRG)	10910	17.3	14000	640	70 x 100

Notes:

a.   The 50% zone is the area in which 50% of the rounds fired will fall.   The first figure mentioned is the width of the zone at right angles to the line of fire. The second figure is the length of the zone along the line of fire.

2.   Note the good range achieved by the rocket system with a comparatively heavy projectile fired from a light equipment.

3.   Note the dimensions of the 50% zone for the rocket system, especially the width of the zone.

SELF TEST QUESTIONS

QUESTION 1   Why are fin-stabilised rockets usually given an amount of spin
as well? How is the spin imparted?

Answer   ....

....

....

....

QUESTION 2   What are the main components of a free-flight rocket system?

Answer   ....

QUESTION 3   Rocket warheads can be much less robust than a comparable
shell casing. Why is this so and what advantage can be
derived from this characteristic?

Answer   ....

....

....

QUESTION 4   What is the main advantage in having a rocket motor with a short
burning time?

Answer   ....

....

....

QUESTION 5   Compared with guns, rockets generally have inferior accuracy
and consistency. List four reasons for this.

Answer   ....

....

....

....

QUESTION 6   State the formula to express the relationship between the
time of travel of a rocket fired from a rail launcher and
the rail length.

Answer   ....

....

....

....

....

QUESTION 7   Briefly describe the two basic types of rocket launcher and
state which one is normally used for FFR.

Answer   ....

....

....

....

....

QUESTION 8   What are the critical dimensions of a convergent-divergent
nozzle?

Answer   ....

....

....

QUESTION 9   Why is solid fuel preferred for FFR?

Answer   ....

....

....

QUESTION 10 Explain the term "platonised propellent".

Answer   ....

....

....

....

....

QUESTION 11   What are the problems in nozzle design for rockets producing
high thrust?

Answer   ....

....

....

QUESTION 12   Explain the term "tip-off".

Answer   ....

....

....

....

ANSWERS ON PAGE 197

8.

Future Trends

INTRODUCTION

In discussing the future one has the advantage of being secure in the knowledge that the opinions proffered are difficult to disprove. Nevertheless the security offered in this regard is far too temporary in the field of artillery to tempt the author to indulge in the business of crystal gazing. Instead, this chapter will be confined mainly to an examination of current trends: trends that are already clearly evident, if not in terms of hardware at least in terms of potential. By necessity, much of the latter will be opinion, albeit hopefully informed. One must, however, bear in mind that regardless of the accuracy with which trends are plotted, the rate of change in world events and advances in technology can alter the groundrules quicker than ever before.

The examination of future trends in this chapter will concentrate firstly on three of the more fundamental aspects of artillery systems: range, calibre and ammunition. The likely effect of automatic data processing in the future will then be discussed, followed by a look at survival of artillery weapons on the battlefield.

RANGE

One often hears the argument that improvements in modern artillery systems have produced maximum ranges that exceed our ability to acquire targets at comparable ranges, the inference being that money would be better spent on improving other components of the overall artillery system. Although one can produce good arguments for giving higher priority to other parts of the system such as protection and rate of fire, it would be a brave gunner who accepts that the present range capabilities are likely to be adequate in the long term. In deciding on the ideal maximum range for a given task the first general criterion to be met is that a weapon system should out-range the comparable system used by the enemy. Unless it fulfils this requirement, commanders will be denied the degree of flexibility in the employment of their artillery resources that will help them achieve a firepower advantage in battle.

Most Western armies are acutely conscious of being outnumbered, both in man-power and equipment. As far as guns, mortars and rockets are concerned, a range advantage is a good way of restoring the balance. It allows artillery re-sources to be protected by employing them in depth beyond the reach of enemy CB, it reduces the need for redeployment, especially in the advance and the with-drawal and gives greater scope for concentrating the fire of a large number of weapons on one target. These are all good, old fashioned reasons for long range but nonetheless still valid for the future, despite the dramatic improvements that have already been made in the last decade. The claim that, at present, artillery weapons cannot always apply their range capabilities to targets in depth has sub-stance; however, remotely-piloted vehicles such as the American AQUILA, pro-mise to improve on the existing methods for locating targets in depth.

The means of increasing range in the future will be two fold. Firstly, there is likely to be a greater use of rockets because of their inherent capacity for greater range for lower equipment weight, compared with guns. Already towed guns are at weights that must be approaching practical limits, while the next generation of SP guns such as SP70 and the Oto Melara 155 mm SP gun will weigh well in ex-cess of 40 tonnes. Recent attempts to improve the range of the M109 155 mm SP to 24 kilometres indicate that such ranges are difficult to achieve with a vehicle weight below 30 tonnes, or more, if the equipment is to remain stable. For some armies strategic mobility requirements will make increases in equipment weight unacceptable and alternative methods of increasing range will have to be used. The alternative methods are all ammunition-based and rely on either some form of post firing boost to the projectile or improvements to the carrying power of the projectile. Rocket assisted projectiles (RAP) have been produced for 105 mm and 155 mm equipments; however there is a penalty of approximately 20 per cent in terms of payload. There are two approaches to improving the carrying power of a projectile. The first is to increase its Ballistic Coefficient (Co). Ballistic Coefficient is the measure of the effect of air resistance on a projectile and can be expressed as follows:

$$ Co \; = \; \frac{M}{x \, \sigma d^2} $$

Where M = mass of the projectile
        d = diameter of the projectile
        x = shape factor
        σ = coefficient of steadiness

The carrying power improves as the value of Co increases. Further, it can be seen from the above expression that a long thin projectile will have good carry-ing power. The problem with long, thin projectiles is that they are more diffi-cult to stabilise and have payload limitations.

The other method of improving the carrying power of the projectile is to use "base bleed". Base bleed projectiles contain a small charge that burns during flight and reduces the drag at the base of the projectile. As with RAP, the use of base bleed incurs a payload penalty; however, the loss in payload is less than for RAP, being in the order of 10 per cent. On the other hand the increase in range for RAP is approximately 25 per cent compared with 15 per cent for base bleed. All of these ammunition-based methods of increasing range are covered in more detail in Volume 3.

CALIBRE

Once a particular calibre and the ballistic solution for it have been accepted there is an understandable reluctance to change. The sheer cost of developing and manufacturing a different calibre system is prohibitive, not to mention the capital investment represented by existing stocks of ammunition of the old calibre. Most countries would also need to consider the desirability of standardisation with allies before taking any unilateral action on the question of calibre. Obviously it will always be difficult to convince allies that the change is worthwhile and if it is worthwhile what the new calibre should be. Sometimes it might be possible to provide guns with replaceable barrels so that stocks of new and old ammunition can be fired.

The UK Light Gun, for example, has been designed to take two barrels: one to fire ABBOT ammunition, the other to fire the old M1 ammunition. Although this method can sometimes be adopted for different ballistic solutions of the same calibre it is far more difficult to "up-gun" an equipment to increase its calibre, because of the changes in the stability equation for the equipment. Although feasible for both towed and SP equipments it would almost certainly not be prac- ticable for towed, primarily because most modern towed equipments are designed to limit maximum weight as much as possible. Consequently any significant in- crease in trunnion pull is likely to necessitate major modifications to the equip- ment.

The effect of becoming committed to a particular calibre may be to perpetuate a system that is becoming increasingly inadequate for its role. Some of the recent attempts to develop new light, towed artillery systems have been discontinued in favour of 155 mm equipments. The French dropped their 105 mm gun project as did the Americans with their 105 mm M204 Soft Recoil gun. In both cases it was as much a question of calibre as anything else. This leaves a situation in which 105 mm calibre is retained on a reduced scale for operations demanding a high degree of strategic and tactical mobility. But already the limitations of 155 mm calibre weapons are also evident.

The basic 155 mm HE projectile was designed for optimum effect against person- nel and lightly protected targets. At present it still maintains its position as the most commonly used nature of projectile. Given the high percentage of armoured targets presented to armies that have opted for 155 mm as the basic calibre, the HE projectile seems less than satisfactory. This is not to say that concentrations of 155 mm HE are totally ineffective: on the contrary, if applied with sufficient weapons and high rates of fire, the desired result can be achieved. Nevertheless, it has already been recognised that for armoured targets, other natures of projec- tile including minelets, terminally guided sub-munitions and cannon launched guided projectiles could be more effective. The effectiveness of all of these pro- jectiles is ultimately limited by calibre. For example, there are finite limits to the cone diameter of shaped charge warheads that can be carried in a 155 mm projectile. It is reasonable to assume that eventually improvements in armoured protection could render 155 mm calibre unsuitable.

Should a larger calibre weapon be required to deliver anti-armour munitions there are several options open. The anti hard target tasks could be left to free flight rocket systems, leaving the rest to 155 mm and mortars. But then it could be said that for some of the remaining tasks 155 mm calibre is still too large. Alternatively the 155 mm systems could be dropped altogether and the range of targets covered by a mixture of mortars and rockets. Clearly there are problems with this solution too and these can be readily identified in Appendix 1 to Chapter 3. A further solution might be to replace 155 mm guns with a larger calibre gun system; however while this may alleviate the hard target problem it would make the guns even less suitable for many close support tasks. The key to this puzzle will be in the improvements that can be made to guns, mortars and rockets but whatever happens it seems highly likely that the future of 155 mm guns could eventually be open to question.

AMMUNITION

Cannon Launched Guided Projectiles

Because of the inadequacies of conventional HE projectiles against tanks the USA has developed a 155 mm projectile that can be guided onto its target. This projectile is called a Cannon Launched Guided Projectile (CLGP) and is now entering service with the US Army. The concept is a result of efforts to improve conventional gun systems against hard targets. It is a first generation system.

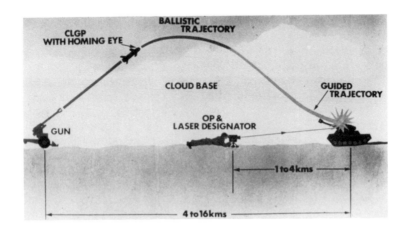

Fig. 79.   CLGP Concept

A CLGP is not a guided missile in the true sense. It is a full calibre projectile fired from a conventional 155 mm gun and follows a normal ballistic trajectory for most of its flight. Towards the end of its trajectory a laser energy seeker in the nose of the projectile homes in on reflected laser energy placed on the target by a laser designator. The outline concept is shown in Fig. 79.

The projectile has three main components: the guidance section containing the seeker, the warhead section, and the stabilisation and control section (see Fig. 80). The seeker has a detector to sense guidance errors. The warhead contains composition B explosive in a shaped charge with a copper liner. A ribbon cable is used to relay command signals from the guidance section to the stabilisation and control section at the tail of the projectile. The stabilisation and control section has two sets of wings that are deployed by centrifugal force when the projectile leaves the muzzle. Additional wings towards the centre of the projectile are deployed at a pre-determined time for the beginning of the guidance sequence. The mid-body wings give the projectile the capability of following either a ballistic or a glide trajectory until the correct laser code is received from the target. Proportional navigation is used until impact.

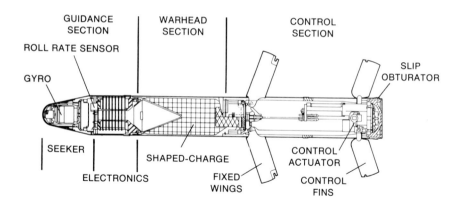

Fig. 80.   CLGP Main Components

The advantages of a CLGP are that in ideal conditions it gives a high probability of defeating a tank with one or two rounds. By comparison up to 2000 conventional HE rounds or 250 improved conventional munitions may be needed to achieve the same result. The advantages in logistics and barrel wear are obvious.

The disadvantages include: the need for some form of target designation; reduced range compared with conventional HE projectiles; the effect of cloud cover in the area of the target; and the reliance on good communications and response. A CLGP has a "footprint" or "area of authority", which is the area, around the point on the ground at which the projectile is directed, within which it can manoeuvre as it approaches the target. The footprint is roughly elliptical in shape and although the projectile can manoeuvre to the edges of the footprint it has a greater chance of a hit at the centre. The hit probabilities vary from about 0.9 at the centre down to 0.2, or less, near the edge. Therefore, it is important that the footprint of the CLGP is superimposed over the target as accurately as possible.

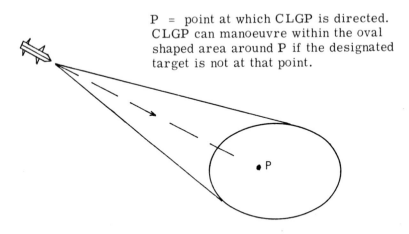

P = point at which CLGP is directed.
CLGP can manoeuvre within the oval shaped area around P if the designated target is not at that point.

Fig. 81.   Footprint for CLGP

The laser target designator must be positioned where it has an uninterrupted line of sight to the target so that the target can be designated during the last 10-15 seconds of its flight. The range of the existing CLGP is less than 15 kilometres and the footprint size varies with range and cloud base height. With a low cloud base at extreme ranges the footprint size can leave little room for manoeuvre. Moving targets, especially if they are single vehicles or groups of widely dispersed vehicles, could be extremely difficult to hit unless the observer designating the target has very good "command" or view of the target area. An illustration of the calculations required for a moving target is shown in Fig. 82.

The direction of motion of the target is shown by the arrows in Fig. 82. On acquiring the target the observer determines its location at A using his laser designator. He does the same at B noting the time taken to travel from A to B and thus determining the velocity and direction of motion of the target. The observer then calculates the point on the ground at which the target and the CLGP will intercept (C in Fig. 82). In determining the location of C the observer extrapolates the velocity and direction data already obtained. He multiplies the velocity of the target by the total time taken to fire the CLGP plus its time of flight to the target. In the example given in Fig. 82 a total time of 200 seconds has been used as a practical yardstick. The time allowed would normally contain a small margin for unforeseen delays, because if the target progresses too far beyond point C it may be outside the footprint of the projectile when it arrives in the target area. Conversely, the observer can withhold fire if the gun is ready to fire too early. Although the observer does not have to keep the target continuously in view throughout its travel from B to C, he must be able to see it during the last 10-15 seconds of the CLGP's flight so that he can illuminate it with his laser designator.

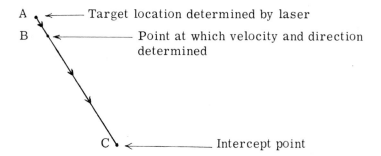

Calculations:

Target displacement measured between A and B using laser = 70 metres

Time interval between A and B = 15 seconds

15 seconds to travel 70 metres .˙. target velocity = 5 m/sec (approx)

Distance B-C: 5 m/sec x 200 sec (reaction time + time of flight) = 1000 metres

Fig. 82.   Moving target calculations for CLGP

Clearly there will be many occasions when the procedure described will be difficult to execute. There are several factors that would possibly cause failure.
These include the necessity for the observer to assume that target direction and
velocity remain constant. In addition, it is apparent that a successful engagement
demands target identification at ranges that may not always be possible and an
uninterrupted view of the target at certain critical points between A and C.
Nevertheless this is not to say that CLGP will be unsuitable for use against moving targets. Ideally CLGP will be used to engage targets in a large array so that
if the target on which his calculations were originally based fails to move as predicted, the possibility exists for another target to be suitably positioned for designation as the CLGP arrives. Furthermore, observers will usually be able to
do some pre-planning for the engagement of both static and moving targets; however, these details will not be covered in this chapter.

The potential of CLGP is such that certain changes in the employment of artillery
and even in the design of equipment may be necessary. Response time is obviously a major consideration in CLGP engagement, therefore.for best results
some guns may have to be reserved for CLGP missions. The weight and length
of existing CLGP exceed that of the normal 155 mm projectile by approximately
35 per cent and 40 per cent respectively. This gives rise to problems of stowage
and automatic loading in SP equipments. Ideally an SP gun tasked for CLGP
missions should have its internal configuration and loading arrangements tailored
to CLGP requirements. Whether this approach would be cost-effective remains

to be seen, since it is still too early for most armies to quantify the true value of
the concept.

Future improvements to CLGP will probably be aimed at reducing the projectile's
length and weight, as well as giving it a capability to guide itself onto targets
without the need for any external designator. Research and development has already begun into extensions of the CLGP concept to include 8 inch calibre projectiles; anti-radiation homing heads for the attack of battlefield radars and jammers;
and extended range guided projectiles with rocket assistance. The efforts in this
field tend to indicate that the CLGP concept is here to stay; however, at the time
of writing the extent of its potential and the likely countermeasures were less
evident.

Propellents

The application of liquid propellents to gun systems has so far been confined to
experimental equipments. Compared with solid propellents, liquid propellents
offer potential for producing increased muzzle velocities and reductions in barrel
heating and wear. Consequently the future use of liquid propellents for guns now
looms as a distinct possibility. There are other inherent advantages too. For
example, with the increased employment of auto-loaders to meet the high rates of
fire demanded in the future any means of reducing the complexity and size of the
auto-loader and improving its reliability will be most welcome.

This consideration is especially important with multi-charge systems firing a
variety of projectiles which of course modern gun systems are. With liquid propellent systems the auto-loader needs only to be able to handle projectiles, with
the propellent being injected into the chamber from a reservoir at the appropriate
stage in the loading sequence. It may also be possible to store the propellent in a
reservoir outside the crew compartment of an SP gun thus releasing more space
for the stowage of projectiles. The metering of propellent into the chamber could
be readily varied to produce a wide variety of ranges for a given elevation, far
more than would be possible with a typical multi-charge system. The effect could
be to increase range coverage and even reduce the time taken for the laying sequence. There are further potential advantages in transport, storage and ease of
replenishment; as well as possible reductions in barrel wear and weapon signature from flash and smoke.

Before one runs away with the idea that liquid propellents could be the answer to
a gunner's prayer the following problem areas should be mentioned. There are
many different options for liquid propellent systems depending on whether the
propellent used is a single component propellent (monopropellent) or a double
component propellent (bipropellent). Monopropellents offer advantages in terms
of ease of handling and low toxicity. Bipropellents offer better performance in
terms of force produced but require a more complex system as the two propellent
components must be metered separately into the chamber. Other possible disadvantages include the fact that a liquid propellent, although safer to use because
it is less susceptible to detonation, is more difficult to ignite. Furthermore, the
ability of liquid propellent systems to match the regularity in muzzle velocities
produced by solid propellents has yet to be proven.

Overall the potential advantages seem to indicate that possible application of liquid propellents to artillery systems will be pursued with much more vigour in the future. At present it is too early to predict that we will eventually see such systems on the battlefield until the pros and cons have been properly quantified.

AUTOMATIC DATA PROCESSING (ADP)

Until recently artillery ADP systems have been restricted to battery level computers. Computers such as FACE (the UK Field Artillery Computing Equipment) and FADAC, the American equivalent, were simply a means of automating the methods of solving gunnery problems. Their introduction did not herald a radical change in the employment of artillery and most aspects of gunnery remained much the same as they were during World War II.

Today there are systems in service or about to enter service that take the application of ADP to artillery much further. These systems include the American TACFIRE system and the French ADP system ATILA, both of which are in service. The British system BATES will be fielded in a few years time. It is difficult to compare these systems but they offer varying degrees of automation that indicate the likely trends for the future.

The ADP systems of the future, of which those mentioned can be regarded as first generation, will have the following potential. Improved ballistics computers will be used to replace the FACE and FADAC generation of equipments, providing faster calculation of gunnery data and more accurate, current ballistic calculations. In addition, distributed processing with local processor stores at command posts will allow autonomous action in the event of system or communication failures. Message handling throughout the system will be automated, relieving operators from tedious tasks presently carried out manually and ensuring the best use of net capacity. Digital communications will be used to link the elements of the system thus reducing net occupancy, providing for a more accurate passage of information, as well as giving greater protection against electronic warfare. Digital communications will also reduce engagement times by decreasing the time taken to pass fire orders. Eventually the communications used may be a mix of voice and digital communications on the one net; however, initially voice and digital nets will be operated in parallel.

Within an artillery ADP system of the future it will be possible to include all types of indirect fire systems from mortars to rockets, as well as other important elements of the system such as locating devices and remotely piloted vehicles. At the extreme ends of the system forward observers will have message entry devices to insert their requirements into the system while at the weapon the possibility exists for the partial or even complete automation of sight setting and laying. Artillery ADP systems with these capabilities will allow for the easy application of artillery resources in ways that hitherto have been either complex and slow or even totally unattainable. In the engagement of targets it will be possible to apply the ideal combination of number and type of weapon, projectile and fuze mix based on instantaneous advice on the optimum solution. The computation of fire plans up to Divisional Level and even higher will be measured in terms of mere minutes including the issuing of orders. Control of resources at the highest level will be much quicker and more flexible than with present means, primarily because of the ability to handle a large number of targets simultaneously and to impose filters in the system to ensure that the priority targets are engaged. As an example the BATES system will be able to handle up to twenty divisional targets at any one time. Command post staff will also be relieved of the routine tasks such as ammunition accounting and recording the locations of units during a fast moving battle. Such information can be automatically entered and stored and instantaneously recalled when required.

These are just a few of the potential advantages of future artillery ADP systems. It is too early to assess their ultimate potential and like all ADP systems their true worth will only be realised after extensive use. One thing seems certain, ADP systems promise to have the greatest impact on artillery since the French first fielded their QF 75 mm gun.

SURVIVAL ON THE BATTLEFIELD

General

The likely threat to artillery weapons on the battlefield of the future will have three main elements, counter battery fire, ground attack and air attack. In this regard the threat in the future is the same as that of the immediate past; however, the changing nature of these elements and the trends in technology that are available to counter them are worthy of consideration.

Counter Battery Fire

All armies regard enemy batteries as high priority targets. It follows, therefore, that the means of avoiding counter battery fire or reducing its effects should also receive high priority. Traditionally the prime methods employed to survive under counter battery fire have been camouflage and concealment; digging; movement; dispersion; and armoured protection. The increased threat from counter battery fire in the future will not be simply a function of the greater number of artillery weapons opposing Western armies, but also a result of the improved munitions and delivery systems that will be employed.

In the past, enemy counter battery fire has come from guns, mortars and, to a lesser extent, free-flight rockets. The projectiles used were usually HE, producing a large number of fragments for optimum effect against personnel and lightly protected vehicles. In the future much greater use will be made of rockets because of their inherent capabilities in terms of range and payload: a trend that has long been apparent in Soviet bloc countries. As mentioned earlier in this chapter, projectiles and warheads used may cease to be predominantly of the HE fragmenting type, with greater use being made of bomblets, minelets, CLGP and terminally guided sub-munitions. All four types will be delivered by rockets and guns; however, as has already been discussed, the use of guns for delivering terminally guided sub-munitions may be limited by calibre. The CLGP and the terminally guided sub-munitions of the future will not require any external designator as is the case with COPPERHEAD, the only existing CLGP. Further-more, the gun launched projectiles and rockets delivering scatterable bomblets and minelets may also be terminally guided so that their payloads can be distribu-ted more accurately at the target end. The overall effect of the use of these im-proved munitions will be a potential for counter battery fire with greatly enhanced accuracy, lethality and area coverage.

The level of armoured protection for SP artillery equipments has thus far been designed to defeat near bursts from HE projectiles. In the future, greater em-phasis will be given to protection against direct hits: especially against hits by shaped charge projectiles and pre-formed fragments. Moreover, the degree of armoured protection against top attack will have to be increased to cope with the steep angles of arrival of guided projectiles and sub-munitions. As an example, it is conceivable that the weight of future SP 155 mm guns could escalate to 55 tonnes or more if they are to have adequate protection.

Frequent movement has long been a means of survival for indirect fire weapons, especially mortars. All indirect fire weapons will be more vulnerable to detec-tion in the future. Although mortars and rockets will still be more easily located than guns, improvements in sound ranging, flash spotting, locating radars, radio direction finding and remotely piloted vehicles will tend to make the difference much less significant. Furthermore, the likely advances in the field of surveil-lance and target acquisition may tend to negate the usefulness of traditional methods of camouflage and concealment: especially after firing has commenced, but even before too. The longer indirect fire weapons remain in action in the same location the more vulnerable they will become. It has always been so; however, as has already been discussed, the lethality and accuracy of counter battery fire will pose a greater risk in the future. The ability to move rapidly and often will, therefore, become increasingly critical. Certainly more impor-tant than was previously the case for gun systems.

Some towed guns already in service are equipped with auxiliary power units to facilitate rapid redeployment over short distances. This trend is likely to con-tinue, as is the trend towards track-mounted mortars. The existing mobility of SP equipments is adequate; however in future the likely weight increases needed to provide armoured protection will demand improvements to engines and sus-pensions if SP systems are to retain their present flexibility in deployment. The likely escalation in the weight of SP guns could degrade their strategic mobility to the extent that armies with a requirement to deploy by air will have to com-promise either on the level of armoured protection or rely on towed guns, mortars and rockets.

If frequent redeployment is to be a successful tactic in the future, a high level of mobility is not the sole requirement. The old conventional methods of providing survey for indirect fire systems will also need to be speeded up so that the total time out of action is kept to a minimum. Traditionally, indirect fire weapons have been dependent on pre-surveyed firing locations. Additionally, on arrival in these locations the weapons required orientation data passed from a theodolite or similar device. The time taken for these preparations is difficult to quantify: the nature of the terrain, the amount of warning given for preparation, together with the degree of dispersion of the weapons in their firing positions all being

contributing factors. Nevertheless, it is reasonable to assume that if weapons
are redeployed 7 or 8 times a day with individual equipments widely dispersed,
the delays in providing survey could be unacceptable. The trend, therefore, will
be towards providing SP equipments with on-board orientation and fixation devices. Certainly such facilities will be almost essential for SP rocket systems
of the future because of their vulnerability to detection and the increasing importance of their role. At the other end of the scale it is less likely that the provision of on-board survey devices will be regarded as cost-effective for tracked
mortars.

The tactic of displacing indirect fire weapons frequently may often be difficult to
accomplish. Lack of deployment areas, insufficient road space in difficult terrain and most important of all, the need to provide continuous fire support will
often mean that weapons may be forced to occupy positions longer than originally
planned. If forced to remain static for long periods the degree of dispersion of
individual equipments becomes much more important as a means of surviving
counter battery fire. Already there are armies that are progressing towards discarding the conventional deployment layouts of 6 or 8 equipments in the area of
approximately 150 m x 200 m in favour of more widely dispersed layouts. The
technology will be available for degrees of dispersion that could result in individual equipments being deployed in widely dispersed locations. In addition to the
on-board orientation and fixation devices already mentioned, each equipment
could conceivably have its own ballistic computer for the production of gun data,
making it completely autonomous: however, the practical problems such as local
defence and logistic support need to be closely assessed. Nevertheless, any
means of achieving greater dispersion to counteract the spread of counter battery
fire concentrations is worthy of consideration, especially with the emergence of
scatterable munitions, both conventional and terminally guided, that can easily
cover the areas occupied by weapon systems deployed to a conventional pattern.

The protection of ammunition delivery means has received scant attention in the
past. Even today most major Western armies deliver ammunition to gun positions in vehicles with little or no ballistic protection. Furthermore, in many
cases the ammunition vehicles to support SP equipments are wheeled and consequently cannot match the mobility of the equipments that they are supporting.
The reason for this is primarily the cost of providing tracked, armoured ammunition vehicles and the desirability of keeping the different types of logistic vehicles
to an absolute minimum. Notwithstanding the reasons for compromise on the
question of ammunition vehicles, the counter battery threat will demand that more
attention be given to their mobility and protection. If we are to ask batteries to
move frequently to avoid counter battery fire, ideally it seems logical to expect
the ammunition delivery means to have a level of mobility to match the weapon.
Similarly, if we are to give SP equipments a high level of ballistic protection, it
follows that a comparable level of protection would be desirable for ammunition
carriers.

Some armies have already recognised the need for mobile, protected ammunition
delivery vehicles. The French Army has a tracked, armoured accompanying
vehicle for its AMX 13 155 mm AMF 3 SP gun; however, it could be argued that
this is only a partial answer to the problem as the accompanying vehicle doubles
as an APC for the crew because the gun is unarmoured and only carries a driver

and a commander when moving (see Fig. 83). The accompanying vehicle has stowage space for a limited amount of ammunition (25 rounds) with a further 30 rounds carried in an unarmoured trailer towed behind the accompanying vehicle.

Fig. 83.   French APC/VCA Tracked

Note: This tracked, artillery accompanying vehicle is
         based on the AMX13 chassis.

At present the US Army uses a tracked, ammunition delivery vehicle, (the M548 load carrier) which has adequate mobility but is unarmoured. A few concepts have been developed in the United States for armoured ammunition resupply vehicles. For example, Bowen-McLaughlin-York have produced an armoured ammunition vehicle based on the M109 155 mm SP gun chassis. The vehicle has aluminium armour and is fitted with a crane. It also features a conveyor belt for transferring ammunition from the vehicle into the M109 SP. The vehicle can carry 100 rounds of 155 mm ammunition and is the obvious answer for integrated ammunition resupply for the M109 system. Clearly this is a great step forward; however, the system of transferring ammunition from the vehicle to the gun may not suit the design characteristics of SP guns of the future, especially those with auto-loaders. Other concepts under development are variations on this theme, being armoured, tracked ammunition vehicles that can back up to an SP gun and replenish it, in some cases even while the ammunition vehicle itself is being re-supplied with additional ammunition.

Fig. 84.   M109 Ammunition Delivery System

Other armies seem less convinced of the need for greater protection and mobility
for ammunition delivery vehicles. The armies of Britain and the Federal
Republic of Germany seem to be fixed on wheeled ammunition vehicles, probably
in the belief that a wheeled vehicle has adequate mobility for the likely area of
operations. Nevertheless, the question of protection goes unanswered. Undoub-
tedly the cost of improved ammunition delivery system is a major consideration
that will influence armies to allocate resources elsewhere; however, the present
trends seem to indicate that integrated ammunition vehicles for SP equipments
will appear on the battlefield in the not too distant future.

Regardless of the trends towards greater mobility and frequent redeployment,
most armies still regard digging as a basic means of protecting weapons and personnel against counter battery fire. The fact that future deployment trends will
leave less opportunity, or even necessity, for digging does not mean that it will
never be done. It remains particularly important for unarmoured equipments
such as ground mounted mortars and towed guns. This is not to say that armoured
SP equipments would not benefit from occupying dug-in positions. Indeed any
position that is to be occupied for a lengthy period should ideally be dug-in. The
armies of France and to a lesser extent the Federal Republic of Germany seem to
be exceptions, opting for a "shoot and scoot" policy. Other armies such as the
US Army and the British Army seem to acknowledge the need for digging even
though their equipment tables seem to fail to reflect the need, obviously on the
assumption that engineer assistance will be provided. Certainly it seems unlikely that in future all SP equipments will be developed with dozer blades for digging although this would be an obvious development. The problem is that even if
SPs were fitted with dozer blades to dig themselves in, it would be imprudent to

employ guns on the other digging tasks that are equally important, such as digging-in ammunition and command posts. A compromise would be to equip sub-units with one or more tractors for digging but the counter argument to this is that to have such equipments organic to the unit would be a waste of resources on the many occasions that they are not required. The use of prefabricated shelters for ammunition and personnel is one avenue that should be further explored.

Local Defence

The threat of attack by infantry and armour on artillery weapon systems will continue; however, it is by no means the greatest threat posed. The threat from counter battery fire is the most serious one and will continue to remain so in the future. Notwithstanding the relative importance of the two types of threat, there are a few trends worth considering in relation to the protection of artillery systems from ground attack. The first is the potential for automation at the weapon and the mechanical handling of ammunition. The advantages offered by these advances are such that there is a temptation to reduce manning to an extent that local defence becomes extremely difficult because of the sheer lack of numbers. One often hears that the advantage of a particular weapon is that it has a reduced manpower requirement. In practice the manpower level for a given sub-unit may be decided by the need for numbers to cope with proper reliefs in a protracted operation and to handle the multitudinous routine tasks associated with local defence. As far as local defence is concerned the alternative is to rely on withdrawal in the face of a ground threat: an alternative that may not always be desirable or even possible.

The techniques likely to be employed to provide protection against counter battery fire have already been discussed. Of these, dispersion and frequent movement will conflict with the needs for strong local defence. Obviously dispersion will increase the number of likely approaches and make early warning and mutual support far more difficult to achieve. There is no easy answer to these problems either in terms of tactics or equipment, although a degree of amelioration may be provided by a greater use of unattended ground sensors and scatterable mines. Frequent redeployment of artillery systems will also exacerbate the local defence problem, because of the lack of time available for camouflage and concealment, digging, and the proper coordination of a defensive layout.

Air Defence

The importance of artillery systems has led to the decision by some armies to provide gunners with equipment for the close air defence of gun areas. The equipment comprises shoulder-fired surface to air missiles, air defence gun systems, as well as purpose designed sights and mounts for the use of small arms in the air defence role. Both the French and the German armies have air defence gun systems integral to artillery units.

The use of small arms in the air defence role, although perhaps good for morale, is of limited value because of poor lethality and low hit probability. The general unsuitability of small arms will increase in the future because of improvements in the performance and protection of aircraft, and the general trend towards smaller calibre small arms ammunition. Shoulder-fired missiles and air defence gun systems offer greater scope for success; however, if employed they will require additional manpower. Moreover, they are both expensive, particularly in the case of a tracked air defence gun system which nowadays costs about 2.5 - 3 times more than a 155 mm SP gun. It could be argued that there may be a case for mounting shoulder-fired missiles on SP guns, or issuing them to towed gun detachments in much the same way as can be done with light or medium anti-tank weapons. The practical constraints in terms of vehicle design, manning and training tend to make the use of such expedients extremely dubious. All things considered, it would appear that the money would be better spent on equipping specialist air defence units to strengthen the protective umbrella they provide. This leaves artillery indirect fire units with no choice but to rely on the same self defence methods against air attack as those already mentioned for defence against counter battery fire.

Some armies have different perceptions of the air defence threat to artillery and may continue to employ or even increase the use of the options already discussed. Financial and manpower resources are the key. Nevertheless, there is a danger that air defence and ground defence of surface to surface artillery resources can become ends in themselves, rather than means of assisting artillery to meet its primary tasks.

END NOTE

The requirement for indirect fire weapons is likely to continue if only because direct fire systems are too easy to neutralise and too inflexible in employment. Guns, mortars and rockets all seem to have an important part to play for the foreseeable future, even though their relative importance may change. That gunners will have to remain adaptable goes without saying. Scarcely a decade has passed this century without important changes in the development of artillery. In this regard the future does not promise to be any different.

SELF TEST QUESTIONS

QUESTION 1   The quest for greater range in artillery systems goes on despite
              the recent advances, especially in 155 mm systems. Why is
              this so?

              Answer   . . . .

                        . . . .

                        . . . .

QUESTION 2   List two of the ammunition-based methods for increasing the
              range of guns.

              Answer   . . . .

                        . . . .

QUESTION 3   Explain the term "Ballistic Coefficient".

              Answer   . . . .

                        . . . .

                        . . . .

                        . . . .

                        . . . .

QUESTION 4   Give reasons why conventional 155 mm HE projectiles may be
              replaced as the basic nature of projectile in the future.

              Answer   . . . .

                        . . . .

                        . . . .

QUESTION 5   What is a CLGP and describe its main characteristics?

              Answer   . . . .

                        . . . .

                        . . . .

                        . . . .

                        . . . .

QUESTION 6 Explain the term "footprint".

Answer   ....

          ....

          ....

QUESTION 7 Now that the first generation of CLGP has appeared, what are
            the likely improvements that will be sought for future CLGP
            and why?

Answer   ....

          ....

          ....

          ....

QUESTION 8 List three potential advantages in the possible use of liquid
            propellents for artillery systems of the future.

Answer   ....

          ....

          ....

QUESTION 9 List four potential advantages offered by the increased use of
            ADP with future artillery systems.

Answer   ....

          ....

          ....

          ....

QUESTION 10 Why will armoured protection for the top of SP equipments
             become more important in the future?

Answer   ....

          ....

QUESTION 11 Give two reasons why some nations are developing tracked,
             armoured ammunition carrying vehicles.

Answer   ....

. . . .

QUESTION 12 How will the counter battery threat of the future affect the
local defence of gun areas?

Answer   . . . .

. . . .

. . . .

. . . .

. . . .

ANSWERS ON PAGE 199

Answers to Self Test Questions

CHAPTER 1

QUESTION 1 — Gustavus Adolphus was one of the first to appreciate the importance of mobility for artillery weapons. He organised his artillery to meet three distinct roles: siege, regimental and field.

QUESTION 2 — The inventor of gunpowder is unknown; however, the English Friar Roger Bacon produced, in the thirteenth century, the earliest record of its ingredients. There are claims that gunpowder existed as a pyrotechnic for fireworks centuries before Bacon's time.

QUESTION 3 — Jean Baptiste de Gribeauval was an Inspector General of Artillery in France during the eighteenth century. His main contributions were in the fields of organisation and logistics for artillery weapons.

QUESTION 4 — The main characteristic that distinguishes a mortar from a gun is that mortars only fire at angles above 800 mils.

QUESTION 5 — Certainly as far as the British were concerned the use of rockets was prompted by the effective employment of rockets against them by Indian gunners at the siege of Seringpatam in 1799. Subsequently, William Congreve was instrumental in developing rockets in England that were used successfully during the first half of the nineteenth century against the French, the Danes and the Americans.

QUESTION 6 — Spin-stabilised rockets were used during the nineteenth century. For example the Hale rocket was spin-stabilised.

QUESTION 7 — The term "Quick Firing" (QF) was originally used to describe a system which was exactly that: at least compared with previous guns. The speed in loading and firing was achieved by the use of a metal cartridge case housing the means of ignition, with the ammunition loaded through the breech and fired from an equipment with an efficient recoil mechanism.

QUESTION 8 — The main reasons for the decline in the use of rockets in the nineteenth century were the arrival of design and manufacturing innovations for guns that made rocket systems far less competitive. These included rifled bores and efficient recoil mechanism.

QUESTION 9 — Tartaglia was an Italian mathematician who produced in his work La Nova Scientia some of the earliest known theories on the shape of trajectories.

185

QUESTION 10   Breech loading originally fell out of favour because the available
               technology for gun manufacture could not solve the problems of
               obturation. Furthermore, once rifled gun bores became a prac-
               tical proposition the difficulties in loading such weapons from the
               muzzle forced gun designers to turn to breech loading.

QUESTION 11   The introduction of the rifled musket made gunners more vul-
               nerable to musket fire at longer ranges than ever before. Be-
               sides inducing them to adopt the rifling principle for their own
               weapons it also gave rise to the importance of indirect fire
               techniques as a means of protecting artillery from direct fire
               weapons.

QUESTION 12   Benjamin Robins was an English mathematician and engineer who
               in the eighteenth century invented the ballistic pendulum and put
               forward some of the basic theories on the stability of projectiles
               in flight.

QUESTION 13   A "trench howitzer" is another name for a trench mortar. It
               originated during World War I.

               CHAPTER 2

QUESTION 1   In selecting the best artillery weapon system for the task, the
               relative importance of weight of projectile, range and accuracy,
               mobility and protection will often vary or conflict. Certain
               primary tasks can usually be identified and to fulfil these tasks
               modern armies are compelled to use more than one type of in-
               direct fire weapon.

QUESTION 2   Recent improvements in methods of providing detection and
               observation at night such as radar, image intensifiers, low light
               television and thermal imaging has tended to reduce the need for
               illuminating ammunition.

QUESTION 3   The main ammunition-based methods of increasing range are to
               improve the carrying power of the projectile or to provide it with
               some form of post-firing boost.

QUESTION 4   The term "range coverage" is used to describe the ability of an
               indirect fire system to fire at different muzzle velocities at the
               same elevation, thus producing different trajectories for that
               elevation.

QUESTION 5   "Accuracy" is a measure of the precision with which the mean
               point of impact (MPI) of a group of rounds can be placed on a
               target. "Consistency" is a measure of the spread of rounds
               about the MPI when the rounds are fired from a gun at the same
               elevation.

QUESTION 6   Gun shields can provide a limited degree of ballistic protection
              for the gun detachment. This could be particularly important if
              the gun is being used in the direct fire role; however, nowadays
              this is less likely. Shields also help protect the detachment from
              blast overpressures. The importance of airportability means
              that the weight of towed guns is kept to a minimum. The trend,
              therefore, is to exclude shields.

QUESTION 7   The term "burst fire" implies a capability to fire a high number
              of rounds in the first 10-20 seconds of an engagement. Burst
              fire can be an effective way of causing casualties before troops
              can take evasive action. The greater the number of casualties
              the more lasting the neutralisation effect.

QUESTION 8   a. Weight of projectile.
              b. Range.
              c. Degree of crew protection provided by the equipment.

QUESTION 9   a. Ballistic protection.
              b. Dispersion.
              c. Camouflage and concealment.
              d. Digging.
              e. Mobility.

CHAPTER 3

QUESTION 1   Yes in calibre 155 mm and above.

QUESTION 2   Advantages

              a. Cost.
              b. Airportability.
              c. Relatively simple.

QUESTION 3   a. Rifled tubes are heavier and more expensive.
              b. Design difficulties if muzzle loading used with a rifled bore.
              c. Ammunition does not need to be as robust.
              d. Higher rate of fire.
              e. A separate firing mechanism not essential.

QUESTION 4   a. Loss of payload.
              b. Accuracy degraded.
              c. Consistency degraded.

QUESTION 5   a. Heavier mortar needed.
              b. Size of fins may have to be unacceptably large.
              c. Spin-stabilised bomb may have to be accepted together with
                  the attendant penalties.

QUESTION 6   Mortars are relatively easy to locate because of their high trajec-
               tory, low velocity and finned bomb. Consequently frequent re-
               deployment is an important means of protection. This, in turn,
               demands a high level of mobility.

QUESTION 7   Guns

               a. Ideally suited to close support tasks because of their good
                   accuracy, consistency, range coverage and their ability to
                   provide sustained rates of fire.
               b. For depth support tasks their comparative poor range, pay-
                   load, area coverage and anti-hard target capabilities make
                   them limited. Extended range and improved conventional
                   ammunition would improve their performance for such tasks.

               Rockets

               a. Comparatively poor characteristics in terms of sustained
                   fire, range coverage, accuracy and consistency make them
                   less suited to close support.
               b. Ideally suited to depth support tasks because of range, pay-
                   load, heavy weight of fire delivered in short period, area
                   coverage and potential for sub-munitions.

QUESTION 8   Guns fire at elevations below 800 mils, mortars do not: this is
               the main difference. However, an orthodox mortar can also be
               distinguished by the fact that it is smooth bore, muzzle loading
               and does not have a recoil mechanism.

               CHAPTER 4

QUESTION 1   The ordnance.

QUESTION 2   Progressive rifling is that form of rifling in which the relation-
               ship between the slope of the grooves and the axis of the bore
               varies, the slope increasing towards the muzzle.

QUESTION 3   Calibre is the diameter of the bore excluding the depth of the
               lands.

QUESTION 4   a. Deep grooves improves degree of guidance given to the pro-
                   jectile in the bore.
               b. Deep grooves reduce the sensitivity of the rifling to wear.
               c. Easier for the driving band to engage in shallow rifling.
               d. Shallower rifling means shallower engraving on the driving
                   band and hence less air resistance in flight.

QUESTION 5   Charge ignites and propellent burns at a rate that increases in
               proportion to the rate of increase in pressure. Shot start pres-
               sure reached and projectile moves. Space available for gases

increases and rate of increase in pressure decreases. Point of
maximum pressure reached when space increase behind the projectile causes pressure loss to equal pressure increase from
burning charge. Projectile continues to accelerate after charge
consumed. Rate of acceleration decreased until retardation
occurs outside muzzle.

QUESTION 6   25-35%.

QUESTION 7   a. Increased chance of muzzle flash.
                b. Greater likelihood of charge not being consumed.
                c. Greater variations in muzzle velocity.
                d. High level of strength needed along a greater length of the
                    barrel.

QUESTION 8   a. Long life.
                b. Strength.
                c. Stiffness.
                d. Suitable mass.
                e. Appropriate centre of gravity.

QUESTION 9   Adequate stiffness so that it does not bend under its own weight.
                It is achieved by selecting a suitable barrel contour.

QUESTION 10   As close as possible to the trunnions.

QUESTION 11   a. Balancing gears.
                b. Counterweights.
                c. Redistribution of mass along length of barrel.

QUESTION 12   a. Wire wound.
                b. Built up.
                c. Composite.
                d. Monobloc.
                e. Loose Barrel/Loose Liner.

QUESTION 13   a. Girder stress.
                b. Radial stress.
                c. Circumferential stress.
                d. Longitudinal stress.
                e. Torsional stress.

QUESTION 14   Torsional stress is generated by the rotation of the projectile as
                it moves up the bore. It produces a twisting effect in the opposite
                direction to the twist of rifling.

QUESTION 15   A method of pre-stressing barrels. Internal pressure is applied
                to the bore and inner layers of metal are stretched beyond their
                elastic limit. Pressure removed and inner layers do not return
                to original shape. Outer layers attempt to return to original
                shape and put inner layers under compression. Two main methods
                are hydraulic method and swage method.

QUESTION 16   a. Cheaper, lower grade steel can be used for a given maximum firing pressure.
               b. Savings in weight.

QUESTION 17   a. Hydraulic.
               b. Swage.

QUESTION 18   Wear caused by the action of hot, high pressure gases in the bore.

QUESTION 19   Wear caused by friction between the projectile and the bore.

QUESTION 20   a. Water needs to be kept in close contact with bore.
               b. Water must be circulated.
               c. A reservoir needed.
               d. Some provision for escape of steam needed.
               e. Space, weight, cost and complexity penalties.

QUESTION 21   Radiating surfaces are often not large enough to dissipate heat.

QUESTION 22   Spontaneous ignition of the charge caused by excessive bore temperature.

QUESTION 23   a. Predicting barrel life due to fatigue.
               b. Establish which steels are best in containing fatigue cracking.

QUESTION 24   a. Withstands rearward pressure on firing.
               b. Houses the firing mechanism.
               c. Provides obturation (BL).
               d. Supports cartridge case in QF system.
               e. Provides method of extraction in QF system.

QUESTION 25   Screw mechanisms.
               Sliding block mechanisms.

QUESTION 26   Loading breech mechanism.

QUESTION 27   It is the means of sealing off the escape of propellent gases. In BL system an obturator pad in a coned seating is secured to the breech screw by the bolt vent axial and the mushroom head. Mushroom head squeezes pad on front of breech screw on firing. Pad expands and forms seal. In QF system the seal is provided by the expansion of the cartridge case on firing.

QUESTION 28   Advantages

               a. Lighter.
               b. No need for cartridge case to provide obturation.

Disadvantages

a. More complex and expensive.
b. Slower in operation.
c. Not as safe to use as a sliding block.

QUESTION 29   Advantage

a. Better distribution of firing loads over the breech ring.

Disadvantages

a. Difficult to machine accurately to ensure that all surfaces
    accept a proportion of the load.
b. More expensive.

QUESTION 30   a. Final movement should throw cartridge case safely clear of
                   breech.
               b. Should not damage cartridge case.
               c. No tendency to bounce when breech is opened.
               d. Cartridge case should not be misaligned from axis of bore
                   as it is being extracted.
               e. Smooth, powerful and relatively slow initial movement to
                   unseat cartridge case.

QUESTION 31   a. Percussion.
               b. Electric.
               c. Percussion and electric.

QUESTION 32   a. Simpler.
               b. Lighter.
               c. More compact.
               d. Quicker response to firing.
               e. Reliable.
               f. Less prone to wear.

QUESTION 33   Their purpose is to prevent fumes from escaping from the bore
               into the crew compartment. Fumes are trapped in a reservoir
               as they pass up the bore behind the projectile. On shot ejection
               the pressure in the bore drops and the fumes trapped in the
               reservoir are expelled back into the bore, through inclined
               ports, in the direction of the muzzle. This action purges the
               chamber and bore of any fumes left behind and forces them out
               through the muzzle.

QUESTION 34   Their purpose is to reduce recoil energy. Gases moving behind
               the projectile strike the muzzle brake baffles, thus exerting a
               force acting in the opposite direction to recoil.

QUESTION 35   Because each set of baffles only deflects about 60% of the gases
               that reach it.

QUESTION 36   Twenty-five times the diameter of the bore. In practice a
               figure of 5 or less is used.

QUESTION 37   a. "Gross Efficiency" is a measure of the percentage reduction
                   in recoil energy to be absorbed by the recoil mechanism as
                   a result of a muzzle brake being fitted.
               b. "Intrinsic Efficiency" is the gross efficiency corrected for
                   the effect of the mass of the muzzle brake.
               c. "Free recoil efficiency" is a measure of the percentage
                   reduction in recoil energy achieved, allowing free recoil
                   during gas action and corrected for the mass of the muzzle
                   brake.

QUESTION 38   a. If retro-fitted there may be a need to fit a counterweight be-
                   hind the trunnions to maintain the balance of the piece within
                   the original design calculations.
               b. Corrections may be needed to muzzle velocity, jump and
                   droop.
               c. Obscuration problems.
               d. Blast overpressures.

QUESTION 39   a. Obscuration problems.
               b. Problems in concealment.
               c. Blast overpressures.

QUESTION 40   a. 2.0 psi for ears.
               b. 20-30 psi for lungs.

QUESTION 41   a. Ear muffs or ear plugs can provide attenuation of up to
                   approximately 35 db.
               b. Shields - limited value.
               c. Baffles to deflect blast - limited value.

CHAPTER 5

QUESTION 1   A carriage fires with its wheels in contact with the ground where-
               as a mounting does not fire with wheels in contact with the
               ground. Mountings can be further classified as mobile mountings
               or SP mountings.

QUESTION 2   The main components of the superstructure are: the saddle; the
               elevating, traversing and balancing gears; the cradle; the re-
               coil system; and the sights.

QUESTION 3   Balancing gears are fitted to counteract any out of balance
               moment caused when the trunnions have not been positioned at
               the centre of gravity of the elevating mass.

QUESTION 4   The cradle supports the barrel and houses the recoil mechanism.
               The cradle is supported by the saddle.

QUESTION 5   "Run-out" is that part of the recoil sequence that commences after the gun has fully recoiled and ends when the barrel has returned to its original position.

QUESTION 6   The main gun design parameters that determine its stability are:

      a.   The weight of the gun.
      b.   The effective trail length.
      c.   The height of the trunnions.
      d.   Trunnion pull.

QUESTION 7   The "foundation figure" is the envelope determined by joining the outermost ground contact points with a straight line. The over-turning moment produced by recoil becomes critical if the line of recoil passes outside the foundation figure.

QUESTION 8

$$R = \frac{WrLr + WsLs}{H}$$

$$= \frac{2000 \times 160 + 5000 \times 115}{45}$$

$$= \frac{597000}{45}$$

$$= 12778$$

QUESTION 9   If a gun is stable at zero elevation and a given recoil length, the recoil length can be shortened at higher elevations without affect-ing stability. The shorter the recoil length the greater the time available to load the ammunition between rounds.

QUESTION 10   If the mass of the recoiling parts is increased less recoil energy will be transferred to them with the result that trunnion pull will be reduced.

QUESTION 11   If the length of recoil is increased there is a decrease in the force acting on the trunnions, energy being the capacity to do work and work being the force applied multiplied by the distance over which it acts.

QUESTION 12   The 2 main components of a recoil mechanism are:

      a.   The buffer.
      b.   The recuperator.

QUESTION 13   A "control to run-out device" is an arrangement that controls run-out by slowing down the movement of the buffer piston as it returns to its original position.

QUESTION 14   "Soft recoil" is the term used to describe the system of recoil
               in which the recoiling parts are held back on a latch under ten-
               sion and released before firing. The charge is ignited while the
               recoiling parts are travelling forward.

QUESTION 15   The advantages are:

               a. Shorter recoil length.
               b. Higher rate of fire.
               c. Reduction in trunnion pull.

               The disadvantages are:

               a. Special arrangements needed to cater for misfires.
               b. The complexity involved in determining the precise moment
                   at which to ignite the charge.

QUESTION 16   Rear mounted trunnions have the following advantages:

               a. Greater percentage of the ordnance is forward of the area
                   used by the detachment in serving the gun.
               b. Easier loading at high angles of elevation.
               c. Reduces the need for variable recoil.
               d. Design for 360° traverse may be easier.
               e. Lower silhouette.

QUESTION 17   A cradle clamp is used to lock the cradle to the basic structure
               to ensure that it remains rigid as the carriage or mounting is
               moved across uneven ground.

QUESTION 18   Limits are imposed on the traverse of some equipments to ensure
               that:

               a. The line of recoil does not pass outside the foundation figure.
               b. The recoiling parts and the traversing mass do not foul the
                   basic structure.

QUESTION 19   "Direct fire" is the term used to describe fire applied when the
               direct vision link between the gun and the target is used to
               engage a target. "Indirect fire" is the term used to describe
               fire applied when the direct vision link is not available or is not
               used. "Laying" is the process of adjusting the gun's line and
               elevation for the engagement of a target.

QUESTION 20   Drift is the lateral deviation of a projectile during flight as a
               result of the spin imparted to it by the rifling in the bore of the
               gun.

QUESTION 21   A reciprocating sight is one that allows for a correction for lack
               of level of trunnions to be applied.

QUESTION 22

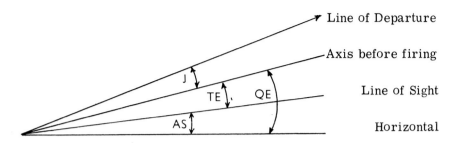

QUESTION 23   "Lead" is the amount of aim-off applied, when aiming at a moving
               target in the direct fire role, to allow for its velocity and direc-
               tion of movement.

QUESTION 24   The main components or groups of assemblies that make up the
               basic structure of a carriage are:

               a.   The saddle support.
               b.   The trails.
               c.   The platform and spades.
               d.   The wheels and axles.
               e.   The suspension and brakes.
               f.   The method of articulation.

QUESTION 25   Pole trails are not used in modern carriages because of their
               limitations in elevation and traverse.

QUESTION 26   Compared with a box trail a split trail has the following
               advantages:

               a.   Greater top traverse.
               b.   Easier access to the breech.

               It has the following disadvantages:

               a.   Longer and heavier trail legs.
               b.   More difficult to reposition for targets beyond top traverse.

QUESTION 27   An equaliser is another name for method of articulation.

QUESTION 28   The variation in the size and shape of spades is the result of the
               differing design requirements for transmitting firing stresses to
               the ground in soils of varying bearing strength.

QUESTION 29   Most gun carriages do not have suspension systems. An example
               of an exception is the UK Light Gun.

QUESTION 30   A mobile mounting is an equipment that may travel on wheels but does not fire off them.

QUESTION 31   Suspension lock-out systems are sometimes used to protect the suspension of an SP gun when the gun fires.

CHAPTER 6

QUESTION 1   a. Smooth bore.
              b. No recoil mechanism.
              c. Restricted to use at elevations above 800 mils.

QUESTION 2   a. Barrel.
              b. Bipod/tripod or monopod.
              c. Baseplate.
              d. Sights.

QUESTION 3   "Bedding-in" is the procedure used to ensure that the baseplate is firmly settled into the ground so that it does not move under firing stresses. Baseplates are bedded-in by firing one or more rounds at a high charge and elevation.

QUESTION 4   Holes are sometimes provided to prevent trapped air from forming an elastic cushion when the mortar is fired and to reduce suction that could make baseplate removal more difficult.

QUESTION 5   Windage is the difference between the diameter of the bore and the diameter of the mortar bomb.

QUESTION 6   Finning is sometimes used to provide a greater surface area for the dissipation of heat. It is far more effective if the barrel is exposed to a breeze. Finning can assist in keeping barrel temperatures down to a workable level.

QUESTION 7   The barrel should be short enough to permit safe loading by a man standing erect. The height of the loader and the length of the bomb are the two main considerations.

QUESTION 8   a. Honing the bore surface to remove any surface irregularities.
              b. Chromium plated bores.

QUESTION 9   a. Bomb can be fired at any time after loading.
              b. Response improved for pre-arranged targets.

QUESTION 10   Rifled mortars

              a. Cost and complexity.
              b. Muzzle loading difficult.
              c. Ammunition more expensive.
              d. Separate trigger mechanism often essential.
              e. Slower rate of fire.

Breech loading mortars

    a. Cost and complexity of system.
    b. Slower rate of fire.
    c. Problems of access to the breech.
    d. Trigger mechanism needed.

QUESTION 11   The main reason is the high cost of purpose-built systems. Hybrid systems will always involve a certain amount of compromise and, therefore, might be less than ideal. Nevertheless, savings in cost may often justify a hybrid solution.

QUESTION 12   Main features/limitations are:

    a. Manportable.
    b. Cheap.
    c. Simple to operate and manufacture.
    d. Short range.
    e. Comparatively inaccurate.
    f. Sometimes without any baseplate or sighting arrangements.

CHAPTER 7

QUESTION 1   To reduce any inaccuracies resulting from thrust misalignment or asymmetric forces produced by fin misalignment.

QUESTION 2   a. The warhead (including the fuze).
    b. The motor (including the combustion chamber and the nozzle.
    c. The launcher.

QUESTION 3   Rockets are not subjected to the high levels of acceleration that are experienced by projectiles fired from guns. The warhead casings can, therefore, be made much thinner, thus providing more space for payload.

QUESTION 4   A short burning time can reduce errors caused by thrust misalignment and surface cross wind effects. In addition, a short burning time producing high accelerations reduces the minimum rail length needed on the launcher.

QUESTION 5   a. Thrust misalignment.
    b. Surface cross wind effects.
    c. Variations in rate of burning of propellent.
    d. Fin misalignment.

QUESTION 6 $$t = \sqrt{\frac{2s}{a}}$$

Where: t = time s = rail length a = acceleration of the rocket

QUESTION 7   a. "Zero length launcher" which is one in which the first motion
                  of the rocket removes it from the restraint of the launcher.
                  Normally used for guided weapons.
              b. "Positive length" or "rail launchers" are long enough to in-
                  fluence the flight of the rocket after it has begun to acceler-
                  ate. The term "rail launcher" encompasses launchers using
                  rail, ramps or tubes.

QUESTION 8   a. The inner slope.
              b. The outer slope.
              c. The diameter of the throat.

QUESTION 9   a. Reliability.
              b. Ease of handling.
              c. Simplicity in design of motor.

QUESTION 10   A platonised propellent is one containing additives to produce a
              more constant rate of burning.

QUESTION 11   a. The critical dimensions (inner slope, outer slope and throat
                  diameter) need to be manufactured to close tolerances.
              b. Nozzle material must have a high melting point and good
                  thermal conductivity.
              c. Nozzle must be robust enough to resist the abrasive action of
                  the gases.

QUESTION 12   "Tip-off" is the tendency of the rocket to tip downwards at launch
              when the forward section of the rocket is unsupported having
              cleared the rail and the rear section is still being guided by the
              rail.

CHAPTER 8

QUESTION 1   Even though the ranges attainable exceed our present ability to
              acquire targets in depth at comparable ranges, long range
              capability remains an important characteristic of artillery
              systems because:

              a. Ideally artillery systems should out-range comparable
                  systems fielded by the enemy.
              b. Long range gives commanders greater flexibility in deploy-
                  ment and employment of resources.
              c. Long range systems can be protected by being sited in depth.
              d. Improvements in long range target acquisition devices can be
                  expected.

QUESTION 2   a. Post firing boost.
              b. Improved ballistic coefficient.

QUESTION 3   Ballistic Coefficient (Co) is a measure of the effect of air
             resistance on a projectile. The carrying power of the projectile
             improves as the value of Co increases.

$$Co = \frac{M}{x \, \sigma \, d^2}$$

             Where: M = mass of projectile
                     d = diameter of projectile
                     x = shape factor
                     σ = coefficient of steadiness

QUESTION 4   155 mm HE projectiles were designed for optimum effect against
             personnel and lightly protected targets. The increasing number
             of hard targets on the battlefield makes the HE projectile less
             than ideal for many engagements and the importance of these
             targets could force armies to use a projectile with a better hard
             target capability as the basic nature of projectile.

QUESTION 5   A CLGP (Cannon Launched Guided Projectile) is a full calibre
             projectile fired from a conventional gun. It follows a normal
             ballistic trajectory for most of its flight but has the capability
             to be guided onto its target in the latter stages of its flight. The
             projectile homes in on reflected laser energy placed on the target
             by a laser designator operated by a forward observer or carried
             by an aircraft or a remotely piloted vehicle.

QUESTION 6   The term "footprint" is used to describe the area on the ground
             within which the projectile can manoeuvre as it approaches the
             target.

QUESTION 7   a.   Greater range.
             b.   No external designator.
             c.   Shorter and lighter projectile.
             d.   More lethal warhead.

QUESTION 8   a.   Increased muzzle velocities.
             b.   Reduction in barrel heating and wear.
             c.   No requirement for charge bags or cartridge cases.
             d.   Potential for improved range coverage.

QUESTION 9   a.   Speed of response improved.
             b.   Simplified fire planning and greater flexibility in the engage-
                  ment of targets.
             c.   Improved data storage and passage of information.
             d.   Digital communications used to link the elements of the
                  system will provide greater security and speed in the
                  passage of information.
             e.   Improvements in speed and accuracy of ballistic computa-
                  tions.

        f.   Less clerical effort required for routine clerical tasks such as ammunition accounting and maintenance of location statements.

QUESTION 10   Armoured protection for the top of SP equipments will become more important with the increased use of guided projectiles and sub-munitions delivered by indirect fire weapons.

QUESTION 11   a.   The need for greater protection against counter battery fire for ammunition supplies co-located with the weapons.

                        b.   The need for ammunition carriers to match the mobility of the weapon systems they are supporting.

QUESTION 12   The counter battery threat will demand the use of increased dispersion, frequent redeployment and rapid digging. Greater dispersion and frequent redeployment both conflict with the needs for the local defence of gun areas.

Glossary of Terms

A

Abbatage

A system of holding a carriage firmly in place by the use of wheel brake shoes dropped under the wheels when the gun is brought into action.

Abrasive Wear

See "Wear".

Accuracy

A measure of the precision with which the MPI of a group of rounds can be placed on the target.

ADP

Automatic Data Processing.

All Burnt

The term used to describe the stage in the sequence of events when a gun is fired at which the propellent is completely consumed.

Angle of Sight

The vertical angle between the horizontal plane through the weapon and the line of sight to the target.

Articulation

The method by which the four points of contact of a split trail carriage maintain contact with the ground when it is uneven.

Autofrettage

A process of pre-stressing a barrel by stretching the inner layers of metal beyond their elastic limit but at the same time stretching the outer layers within their elastic limit. The effect is that the outer layers compress the inner layers. There are two approaches to this process: "Hydraulic Autofrettage" and "Swage Autofrettage".

Axis of the Bore

The straight line passing through the centre of the bore.

B

Balancing Gear

A device used to counteract the out-of-balance movement present when the trunnions are not located at the centre of gravity of the elevating mass. "Equilibrator" is another term used for balancing gear.

Ballista

An ancient, mechanical, torsion-powered weapon for hurling missiles.

Ballistic Coefficient

A measure of a projectile's carrying power. It is a function of the projectile's mass, diameter, shape, and coefficient of steadiness.

Ballistics

The study of the motion of projectiles.

Basic Structure

That part of a gun in contact with the ground. It transmits the firing stresses to the ground and supports the "superstructure".

BATES

Battlefield Artillery Target Engagement System: the British Army's artillery ADP system of the future.

Bedding-In

The process of fixing a mortar baseplate firmly into the ground by firing to ensure that it does not rock during subsequent firing.

BL

Breech Loading (See "Obturation").

Bore

The interior of a barrel from the rear of the chamber to the muzzle.

Breech

A mechanism used to close the end of a gun when the projectile is loaded and to withstand the rearward pressure of propellent gases when the gun is fired. In QF equipments it supports and extracts the cartridge case. In BL equipments it provides the means of obturation. It also carries the firing mechanism.

Buffer

A cylinder containing oil, springs, gas or a combination of these for the purpose of controlling and arresting the recoiling parts during recoil.

Built-Up Barrel

A type of barrel constructed by shrinking two or more concentric tubes in such a way that the inner tube/s is pre-stressed.

Burst Fire
> A high rate in fire produced in a short period. For 155 mm equip-
> ments a rate of 3 rounds in 15 seconds or less is usually described
> as a burst fire rate.

C

Calibrating Sight
> A sight that automatically compensates for a gain or loss of MV from
> the range table standard MV.

Calibre
> The diameter of the bore excluding the depth of the lands.

Calibre Length
> The bore length expressed in calibres.

Cannon Launched Guided Projectile (CLGP)
> A full calibre projectile fired from a conventional gun/howitzer that
> has some form of terminal guidance although for most of its trajec-
> tory it follows a normal ballistic trajectory.

Carriage
> An equipment that fires with its wheels in contact with the ground.

Chamber
> The smooth portion at the breech end of the ordnance shaped to
> accommodate the charge.

CLGP
> See "Cannon Launched Guided Projectile".

Commencement of Rifling
> The point in the bore where the grooves reach maximum depth.

Compensating Sight
> A sight that compensates for drift.

Consistency
> The measure of the degree of spread of a group of rounds about their
> MPI. The smaller the spread the better the consistency.

Control to Run-Out
> The means or method of controlling the final stages of run-out to
> prevent the recoiling parts from jolting back into their original
> position before firing.

Convergent-Divergent Nozzle
> The commonly used type of rocket nozzle used to change heat and
> pressure energy into kinetic energy. Sometimes called a "De Laval
> Nozzle."

COPPERHEAD
> The name given to the first CLGP. Produced in the USA by Martin Marietta for 155 mm weapon systems.

Cradle
> The component of the superstructure that carries the ordnance and enables it to slide axially on recoil. It is pivoted about the saddle to allow the ordnance to be elevated and depressed.

Cradle Clamp
> An arrangement to prevent damage to the elevating gear by preventing movement between the saddle and the elevating mass when the gun is on the move.

Cut-Off Gear
> A means of shortening length of recoil when the gun is elevated.

D

De Laval Nozzle
> See "Convergent-Divergent Nozzle".

Demiculverin
> An early form of muzzle loaded artillery.

Depth of Rifling
> The measurement of rifling depth in the bore of a gun taken from the top of a land to the bottom of the groove.

Direct Fire
> Fire in which the weapon is directed at the target using the direct vision link between the sight and the target.

Drag
> Resistance to the motion of a projectile through the air caused by the region of low pressure behind the base of the projectile: sometimes called "base drag".

Drift
> The lateral deviation of a projectile in flight resulting from the spin imparted to it by the rifling of a gun.

Driving Band
> A band (usually soft metal) fitted around a projectile for the purpose of engaging in the rifling.

E

Elastic Limit
The stress limit for elastic as opposed to plastic deformation of metals.

Elevating Gear
The gear arrangement that controls movement in the vertical plane of the elevating mass about the trunnions.

Elevating Mass
All parts of the equipment which elevate. In most cases it includes the ordnance, the cradle and the recoil system.

Elevation
The vertical (acute) angle between the horizontal plane and the axis of the bore.

Equilibrator
See "Balancing Gear".

Erosive Scoring
The localised removal or scoring of metal from the bore surface caused by an imperfect seal of the driving band allowing high pressure gas escape forward. It most frequently develops on the upper surface of the bore close to the commencement of rifling.

Erosive Wear
See "Wear".

Extraction
The removal of the cartridge case or the vent tube from the breech.

Extractors
Mechanical devices for removing cartridge cases from the breech. They may also serve to hold sliding block breech mechanisms in the open position.

F

FACE
Field Artillery Computing Equipment: the British artillery battery-level ballistic computer.

FFR
Free Flight Rocket.

Finning
The raised, grooved surface found on some barrels, especially mortar barrels, for the purpose of improving heat dissipation by increasing the external surface area of the barrel.

Fixation
> The process of determining the coordinates of a weapon system, an observation post, or a target.

Firing Lock
> That part of a firing mechanism that carries the striker.

Firing Mechanism
> A device for initiating the ammunition primer.

Firing Post
> A term used to describe a fixed firing fin fitted into the base of a mortar barrel.

Firing Table
> See "Range Table".

Foundation Figure
> The envelope of the axis of overturning in the horizontal plane.

Free Recoil Efficiency
> A measure of the percentage reduction in recoil energy resulting from the use of a muzzle brake. It allows for free recoil during gas action and is corrected for the mass of the muzzle brake. (See also "Gross Efficiency" and "Intrinsic Efficiency").

Fume Extractor
> An attachment to the barrel of an SP gun which ensures that propellent fumes do not escape into the crew compartment when the breech is opened.

G

g
> gram.

GAP
> See "Gun Aiming Point".

Girder Stress
> The stress produced by the bending of a barrel by virtue of its weight and length.

Gross Efficiency
> A measure of the percentage reduction in recoil energy as a result of a muzzle brake being fitted. It does not take into account the mass of the muzzle brake (see also "Free Recoil Efficiency" and "Intrinsic Efficiency").

Gun

> Originally this term was reserved for comparatively long range equipments firing a relatively small shell, usually a fixed charge, at a high MV and at a low trajectory. Nowadays the term is often used as a generic term for both guns and howitzers, although in some cases the distinction is maintained and in other cases when the equipment meets the characteristics of a gun and a howitzer the term "gun/howitzer" is adopted (see also "Howitzer").

Gun Aiming Point

> A reference point, real or artificial, used during the engagement of targets by indirect fire to ensure that the weapon is pointing in the right direction.

H

HE

> High Explosive.

High Angle Fire

> Fire delivered by indirect fire weapons at elevations above that at which the maximum range for which the equipment is achieved. In general terms it is fire above elevations of 800 mils.

Howitzer

> A comparatively short range equipment firing a relatively heavy projectile at a low muzzle velocity and using variable charges (see also "Gun").

Hydraulic Autofrettage

> A process for pre-stressing gun barrels by the application of high fluid pressure (see also "Autofrettage" and "Swage Autofrettage").

I

ICM

> Improved Conventional Munition.

in

> inch.

Indirect Fire

> Fire delivered at a target without the use of a direct vision link between the weapon sight and the target.

Intrinsic Efficiency

> A measure of the percentage reduction in the recoil energy to be absorbed by the recoil system as a result of a muzzle brake being fitted and corrected for the effect of the mass of the muzzle brake (see also "Free Recoil Efficiency" and "Gross Efficiency").

J

Jump
> The angle between the axis of the gun when laid and the line of departure.

L

Lands
> The raised portions between the grooves of the rifling in the bore.

Laying
> The process of adjusting the gun for line and elevation.

LBM
> Lever Breech Mechanism.

Lead
> The aim-off applied to the sights to allow for the lateral movement of the target during the time of flight in direct fire engagements.

Line
> Direction or azimuth when used to describe the process of laying.

Line of Departure
> The direction of motion of a projectile as it leaves the muzzle.

Line of Sight
> The line between the sight and the target.

Lock
> The part of the firing mechanism that carries the striker. Sometimes referred to as a firing lock.

Longitudinal Stress
> Stress in a gun barrel resulting from the action of the driving band and the difference in gas pressure between the front and the rear of the driving band. It causes a localised stretching effect along the length of the barrel as the projectile moves through the bore.

Loose Barrel
> A type of barrel construction that incorporates a jacket over the highly stressed parts of the barrel to provide additional support. The barrel can be removed for replacement, or for stripping in the case of pack equipments.

Loose Liner
> An earlier version of the loose barrel concept the difference being that the jacket extends along the length of the barrel.

Low Angle Fire

> Fire delivered at elevations below that at which the maximum range
> for the equipment is achieved. In general terms it is fire below 800
> mils.

M

Magnus Effect

> One of the components of the drift of projectiles that causes projec-
> tiles rotating with right hand spin to drift to the left if the nose of the
> projectile is above the trajectory or to the right if it is below the
> trajectory. It is caused by the build-up of airpressure on one side of
> the projectile.

Mean Point of Impact

> The centre point of the spread of a group of rounds fired from a gun
> at the same line and elevation.

Metal-to-Metal Obturation

> A method of obturation used in some BL systems in which an insert
> in a sliding block makes close contact with a metal ring or bush moun-
> ted in the chamber face of the barrel. The advantage of this method
> is that a sliding block mechanism can be used with a BL ordnance
> (see also "Obturation" and "Obturator").

Mil

> A unit of angular measurement. One degree = 17.8 mils.

MLRS

> Multiple Launcher Rocket System.

Monobloc Barrel

> An equipment that fires without its wheels in contact with the ground.

Mortar

> An indirect fire weapon which, in its conventional form is muzzle
> loaded, has a smooth bore, transmits firing stresses directly to the
> ground and only fires in high angle. Unorthodox versions using
> rifled bores and/or breech loading do exist. Some mortars also
> have a low angle capability.

MPI

> Mean Point of Impact.

Muzzle Brake

> An attachment to the muzzle of a barrel for the purpose of reducing
> the rearward momentum of the recoiling parts by deflecting sideways
> some of the gases emerging from the bore after shot ejection.

Muzzle Preponderance
> The term used to describe the unequal weight distribution of a barrel with rear trunnions in relation to its centre of gravity. Barrels with rear trunnions are sometimes said to be "muzzle heavy" which is another term used to describe the same condition.

Muzzle Velocity
> The velocity with which the projectile leaves the muzzle.

MV
> Muzzle Velocity.

N

Non-Rigidity of Trajectory
> The term used to describe the fact that, with indirect fire, targets at the same horizontal range but at different elevations must be engaged with projectiles following different trajectories.

Nozzle
> A component in the base of a rocket that changes the heat and pressure energy of escaping propellent gases into kinetic energy. The gases are expanded in the nozzle to lower temperatures and pressures, reaching a very high velocity in the process. The rate of change of momentum of the gases provides the propulsion for the rocket.

O

Obturation
> The prevention of the escape of gases produced by the charge and the means of ignition, through the breech. There are two approaches to obturation:
> a.   QF Obturation - in which the cartridge case provides the means of obturation.
> b.   BL Obturation - in which the means of obturation is provided by a resilient pad which seals the rear of the chamber.

Obturator
> A resilient pad which fits in a coned seating at the rear end of the chamber of BL systems. On firing the pad is compressed and expands to prevent the rearward escape of gases (see also "Metal-to-Metal Obturation").

Open Jaw
> A type of breech ring which is slotted in the rear face to accept the block.

Ordnance
> The term used to describe the group of components or assemblies in

a gun comprising the barrel and its attachments, the breech and the firing mechanism. It is also used as a general term to describe guns, howitzers and mortars.

Orientation

The process of initially directing guns at the required bearing for the subsequent engagement of targets.

Out of Balance Moment

The moment caused by the displacement of the centre of gravity of the elevating mass from the trunnions. Its magnitude varies with the weight of the elevating mass the distance from the centre of gravity to the trunnions and the angle of elevation.

P

Pad Obturator

See "Obturator".

Platforms

The main part of the basic structure of some mobile mountings to which are attached the stabilising girders and wheels.

Platonisation

A means of achieving a constant rate of burning over a specific range of pressures by the addition of lead salts to the propellent.

Pounder

In the past some equipments were described by the weight of projectile fired. For example the 25 Pounder fired a projectile weighing 25 pounds. Nowadays the trend is to describe ordnance by their calibre expressed in milimetres, inches, or sometimes centimetres.

Primer

A component in the base of a cartridge case. It contains a cap for initiation and an ignition charge for the propellent. It functions either by percussion or electrical means.

Projectile

A missile fired from a piece of ordnance.

psi

Pounds per Square Inch.

Q

QF

Quick Firing (see "Obturation").

Quadrant Elevation
> The elevation at which a gun is laid to achieve the desired range under the prevailing conditions.

R

Radial Stress
> The stress acting outwards on the walls of the barrel produced by the gas pressure generated by the burning propellent in the bore.

Range
> The distance in metres between gun and target.

Range Table
> A compilation of data on the performance of a gun produced at firing trials and used as the basis for the calculation of gun data for the engagement of targets. Sometimes called a "Firing Table".

RAP
> Rocket Assisted Projectile.

Rapson Nut and Screw
> A type of traversing gear employing a recirculatory ball race.

Reciprocating Sight
> A sight that allows for lack of level of trunnions.

Recoil
> The rearward movement of the ordnance on firing, relative to the mounting or carriage.

Recoil Cycle
> The term used to describe the rearward movement of the ordnance (recoil) and the return of the recoiling parts to their original position (run-out).

Recuperator
> The component of the recoil mechanism that returns the recoiling parts to their run-out position and holds them there at all angles of elevation.

Remaining Velocity
> The velocity of the projectile at the point of impact.

Rifling
> The system of helical grooves cut into the bore for the purpose of imparting spin to the projectile.

Rocking Bar

A type of sight that enables an elevation to be set on the elevation scale which in turn tilts the sight bracket through the same angle. Sight and ordnance are elevated together until the bubble on the sight bracket is level and the desired elevation is thus applied to the ordnance.

Run-Out

The return of the recoiling parts, after recoil, to their original positions.

S

SADARM

Sense and Destroy ARMour: a terminally-guided sub munition.

Saddle

The component of a gun's superstructure that carries the elevating mass but does not elevate with it. It traverses on the basic structure and is connected to it by some form of holding down arrangement.

Saddle Support

The part of the basic structure that supports the saddle. The trails and the wheels are attached to it.

Screw Breech

A type of breech mechanism that employs mating threaded surfaces on the breech ring and the breech block (screw). Sometimes called a breech screw mechanism.

Shaped Charge

A type of warhead with its explosive charge shaped in such a way that the explosive energy is focussed into a powerful and highly penetrating jet. The effect is sometimes called the hollow charge effect, the Newmann effect or the Monroe effect.

Shot Ejection

The point or moment in time at which the projectile leaves the bore.

Shot Seating

The conical portion of the bore where the chamber diameter is reduced to join the rifled portion.

Shot Start Pressure

The level of bore pressure at which the projectile begins to move.

Shot Travel

The distance along the bore between the base of the projectile when loaded, to the muzzle.

Sight

> The instrument used to direct ordnance in the required bearing and/or range.

Sliding Block

> A type of breech mechanism mainly used in QF equipments but has been used for BL equipments. Its main components are the breech ring (either open or tied jaw) and a sliding block to close off the breech. The block can be designed to slide either vertically or horizontally.

Slipper

> A gun component that connects the barrel to the cradle and the recoil system. It includes arrangements to prevent the barrel from turning and also takes the longitudinal thrust on firing.

Soft Recoil

> A system of recoil that requires the recoiling parts to be held back under tension on a latch before firing. On firing the recoiling parts are released, before charge ignition, so that their forward momentum reduces the amount of recoil.

Sole Plate

> Part of the basic structure of some mobile mountings. It provides a contact point with the ground and permits a limited top traverse.

SP

> Self Propelled.

Spade

> An attachment to the trail of a gun to prevent movement to the rear when the gun is fired.

Stabilising Girder

> The component of a mobile mounting that performs a similar function to the trail of a carriage.

Stability

> Mountings and carriages are considered stable if the main supports remain stationary during the recoil cycle.

Striker

> The component of a firing mechanism that initiates the primer.

Superstructure

> The group of components in a carriage or mounting that provides support for the ordnance.

Swage Autofrettage

> A method of autofrettage in which an oversize mandrel or swage is forced through the bore to overstrain the inner layers of metal. The

advantage of this method is that it can be used to pre-stress barrels at pressures that are unattainable by hydraulic autofrettage (see "Autofrettage").

T

Tangent Elevation
The vertical (acute) angle measured from the line of sight to the target and the axis of the bore.

TGSM
<u>T</u>erminally <u>G</u>uided <u>S</u>ub <u>M</u>unition.

Thrust Misalignment
A contributing factor to inaccuracy in rocket systems arising from any displacement between the axis of thrust and the centre of gravity of the rocket.

Tied Jaw
A type of breech ring used in sliding block mechanisms. The breech block recess is cut transversely through the ring, the rear face of the ring being cut away only sufficiently to enable loading. Inherently stronger than open jaw type but more difficult to manufacture.

Time of Flight
The time taken for the projectile to reach the target after the gun is fired.

Tip-Off
The tendency of the nose of a rocket to tip downwards on leaving the launcher rail. It is prevented by launcher design that permits the simultaneous release of front and rear rocket holding points.

Torsional Stress
A twisting effect within the barrel wall caused by the rotation of the projectile as it proceeds up the bore.

Trails
The components of the basic structure of a carriage that transmit firing stresses to the ground, keep the weapon steady in the firing position and connect it to the prime mover.

Trajectory
The curve described by the centre of gravity of a projectile during its flight.

Traversing Gear
The gear that controls movement in the horizontal plane between the saddle and the saddle support.

Trebuchet

> An ancient mechanical artillery weapon that used a counterweight to hurl missiles.

Trench Howitzer

> A name used for mortars during World War I (see "Mortar").

Trough Cradle

> A type of cradle, shaped like a trough, which houses the recoil mechanism.

Trunnion Pull

> The force at the trunnions parallel to the axis of the bore, less the component of the mass of the recoiling parts.

Trunnions

> The point about which a gun elevates.

Twist of Rifling

> The length of bore, expressed in calibres, in which the rifling helix describes one revolution. The twist can be either uniform or varying.

V

Vent Tube

> The ammunition component used in BL charge systems to initiate the igniter bag attached to the propellent charge.

W

Wear

> There are two main types of wear in barrels: "erosive wear" which is the removal of metal from the bore surface by gas action and "abrasive wear" which is the removal of metal from the bore surface by mechanical friction between the projectile and the bore.

Welin Breech

> A type of screw breech mechanism with interrupted threads on stepped segments. The threads are cut on varying radii to achieve a large bearing surface although only a small turning movement is needed to screw the breech home.

Windage

> The difference between the diameter of the bore of a mortar and the diameter of the bomb.

Wire Would Barrel

> A type of barrel construction in which pre-stressing is achieved by winding wire under tension on an inner tube.

WP

White Phosphorus.

Y

Yaw

The angle between the axis of the projectile and the tangent to the trajectory.

Z

Zero Length Launcher

A type of rocket launcher in which the first motion of the rocket removes it completely from the constraint of the launcher.

Bibliography

Baker D., "The Rocket", New Cavendish Books, London, 1978.

"Ballistics and Gunnery", HMSO, London, 1938.

Batchelor J., Hogg I., "Artillery", Macdonald, London, 1972.

Bethall H.A., "Modern Guns and Gunnery", Macmillan & Co., London, 1909.

Bragg S.L., "Rocket Engines", Newnes, N.Y., 1962.

Bruce G., "Harbottles Dictionary of Battles", Granada, London, 1979.

Chant C. (Ed), "How Weapons Work", Marshall Cavendish Ltd., Hong Kong, 1980.

Daish C.B., "The Drift of Projectiles", RMCS, Shrivenham, 1971.

Hayes T.J., "Elements of Ordnance", Chapman and Hall, London, 1938.

Hogg I.V., "The Guns 1914-18", Ballantine, New York, 1972.

"The Guns 1939-45", Ballantine, New York, 1970.

"Grenades and Mortars", Ballantine, New York, 1974.

Hogg O.F.G., "Artillery : Its Origin Heyday and Decline", Hurst, London, 1970.

"Jane's Infantry Weapons 1980-81", Jane's Publishing Co., London, 1980.

"Jane's Armour and Artillery 1979-80", Jane's Publishing Co., London, 1979.

"Jane's Weapon Systems 1980-81", Jane's Publishing Co., London, 1980.

"Journal of the Royal Artillery", RA Institution, Woolwich, London (various copies).

Robins B., "New Principles of Gunnery", Richmond Publishing Co., Richmond, 1972.

Rogers H.C.B., "A History of Artillery", Citadel Press, N.J., 1975.

"Treatise on Military Carriages", HMSO, London, 1911.

"Treatise on the Construction of Ordnance", HMSO, London, 1879.

Wilson A.W., "The Story of the Gun", RA Institution, Woolwich, London, 1968.

Zucrow M. J., "Aircraft and Missile Propulsion", Chapman and Hall, London, 1979.

The above list is not a compilation of all references consulted, as much of the information presented in this book has been obtained from previously unpublished sources and personal notes gathered over the years. The references mentioned should serve as a good guide to those interested in further details on guns, mortars and rockets.

Index